'The very best social science always exposes a new idea and inaugurates original lines of enquiry leaving us wondering why we had never thought of it before. This volume inaugurates a very original approach to understanding the problem of time in societies emerging from conflict. Transitional justice scholars, specialists in peace processes and in victim studies can learn a great deal from the centrality given to the phenomenon of time in the volume. Transitional time, as we might call it, is theorised and discussed empirically in a rich array of case studies, making the volume essential reading for those whose eyes have now been opened to what was so obvious that we overlooked it in the past.'

John D Brewer, Professor of Post Conflict Studies, Queen's University Belfast. Author of *Peace Processes: A Sociological Approach*

'With detailed case studies from Africa, Asia, Latin America, and the Middle East, this impressive collection puts time and temporality at the front and center not only of studies of peacebuilding and transitional justice but also, and as importantly, of sociological inquiry at large.'

Javier Auyero, Professor of Sociology, University of Texas, Austin. Author of *Patients of the State. The Politics of Waiting in Argentina*

'Few will contest that issues of time and temporality are central to transitional justice and peace building. However, until recently these issues have received surprisingly little scholarly attention. This book brings a very welcome change to this situation. The volume offers a very rich and innovative collection of essays that are highly relevant for both theoretical analysis and policy making. Strongly recommended to all interested in transitional justice and conflict resolution as well as the sociology or ethnography of time.'

Berber Bevernage, Professor of Historical Theory, Ghent University, Belgium. Author of *History, Memory, and State-Sponsored Violence: Time and Justice*

Time and Temporality in Transitional and Post-Conflict Societies

Implicit conceptions of time associated with progress and linearity have influenced scholars and practitioners in the fields of transitional justice and peacebuilding, but time and temporality have rarely been systematically considered.

Time and Temporality in Transitional and Post-Conflict Societies examines how time is experienced, constructed and used in transitional and post-conflict societies. This collection critically questions linear, transitional justice time and highlights the different temporalities that exist at local and institutional levels through original empirical research.

Presenting empirical and often ethnographic research from Argentina, Brazil, Colombia, Cambodia, Mozambique, Palestine/Israel, Rwanda and South Africa, contributors use a temporal lens to investigate key issues including: transitional justice institutions, peace processes, victimhood, perpetrators, accountability, reparations, forgiveness, reconciliation and memorialisation.

This timely monograph will appeal to undergraduate and postgraduate students, as well as postdoctoral researchers, interested in fields such as political science, international relations, anthropology, transitional justice and conflict resolution. It will also be relevant to conflict resolution and peacebuilding practitioners.

Natascha Mueller-Hirth is a Lecturer in Sociology at Robert Gordon University, Aberdeen, UK.

Sandra Rios Oyola is a Lecturer in International Studies at Leiden University, The Hague, Netherlands.

Routledge Advances in Sociology

For a full list of titles in this series, please visit www.routledge.com/series/SE0511

238 Reflections on Knowledge, Learning and Social Movements
History's Schools
Edited by Aziz Choudry and Salim Vally

239 Social Generativity
A relational paradigm for social change
Edited by Mauro Magatti

240 The Live Art of Sociology
Cath Lambert

241 Video Games as Culture
Considering the Role and Importance of Video Games in Contemporary Society
Daniel Muriel and Garry Crawford

242 The Sociology of Central Asian Youth
Choice, Constraint, Risk
Mohd. Aslam Bhat

243 Indigenous Knowledge Production
Navigating Humanity within a Western World
Marcus Woolombi Waters

244 Time and Temporality in Transitional and Post-Conflict Societies
Edited by Natascha Mueller-Hirth and Sandra Rios Oyola

245 Practicing Art/Science
Experiments in an Emerging Field
Edited by Philippe Sormani, Guelfo Carbone and Priska Gisler

Time and Temporality in Transitional and Post-Conflict Societies

Edited by
Natascha Mueller-Hirth and
Sandra Rios Oyola

Routledge
Taylor & Francis Group
LONDON AND NEW YORK

First published 2018byRoutledge

2 Park Square, Milton Park, Abingdon, Oxfordshire OX14 4RN
52 Vanderbilt Avenue, New York, NY 10017

Routledge is an imprint of the Taylor & Francis Group, an informa business

First issued in paperback 2019

British Library Cataloguing in Publication Data
A catalogue record for this book is available from the British Library

Library of Congress Cataloging in Publication Data
A catalog record has been requested for this book

ISBN: 978-1-138-63136-6 (hbk)
ISBN: 978-0-367-43181-5 (pbk)

Typeset in Times New Roman
by Taylor & Francis Books

Contents

List of figures

Acknowledgements

Our biggest thank you goes to the contributors to this book: Richard Benda, Luisa Gandolfo, Alexander Laban Hinton, Gisele Iecker de Almeida, Victor Igreja, Eva van Roekel and Valérie Rosoux. We are very grateful to you for your excellent and thought-provoking work, for sharing our enthusiasm for this project, and for complying with our various requests over the last year. It has been a real pleasure to work with you and to jointly produce this collection.

Thank you very much to Elena Chiu and Emily Briggs at Routledge for their support, patience and professionalism.

Thank you to Routledge for permission to reproduce Alexander Laban Hinton's essay 'Transitional justice time: Uncle San, Aunty Yan, and outreach at the Khmer Rouge Tribunal'.

Natascha Mueller-Hirth would like to thank all the participants in her research for sharing their experiences, which informed not only her chapter but the focus of the collection as a whole, and her lovely and supportive colleagues in the School of Applied Social Studies at Robert Gordon University. Thank you to Adam Doran for everything, always, and to Mandisa and Finn.

Sandra Rios Oyola would like to thank her colleagues at International Studies and Latin American Studies at Leiden University for their support and collegiality. Thank you Tijs Zwinkels for your caring companionship and to Victoria Guerrero for your good humour.

For both of us, one of the real joys of this project has been the opportunity to work together with our colleague and friend; our many conversations and the revisions of each other's work were essential in helping us to formulate our thoughts in a way that hopefully does some justice to the depth of the questions we have explored.

Contributors

Richard M. Benda is a Rwandan academic and researcher. Currently, he is a Tutor and a Research Fellow at Luther King Education Trust/University of Manchester where he teaches Contextual Theology and Research Methods. He holds a BA (2007), MA (2008) and a PhD (2013) in Religious Philosophy and Political Life (University of Manchester). He also holds a master's degree in Law from Kigali Independent University (2002), where he taught Criminal Law and Constitutional Law (2003–2004). His doctoral thesis, *The Test of Faith: Christians and Muslims in the Rwandan Genocide*, is in the process of publication. His postdoctoral research interests are genocide studies, post-conflict transitional theories, transgenerational oral histories and practical theology.

Luisa Gandolfo is based in the Department of Sociology at the University of Aberdeen, Scotland. She studied Arab and Islamic Studies at the University of Exeter before completing a postdoctoral fellowship at Durham University, and a EURIAS-NIAS Fellowship at the Netherlands Institute for Advanced Study (2014–2015). Luisa's research focuses on post-war reconciliation, memory, and socio-cultural practices in the Middle East and North Africa, with particular foci on Palestine-Israel and Tunisia, and she is the author of *Palestinians in Jordan: The Politics of Identity* (2012).

Alexander Laban Hinton is Director of the Center for the Study of Genocide and Human Rights and Professor of Anthropology and Global Affairs at Rutgers University, Newark. He holds the UNESCO Chair in Genocide Prevention and served as an expert witness at the Khmer Rouge Tribunal in 2016. He is the author of *Man or Monster? The Trial of a Khmer Rouge Torturer* (2016), the award-winning *Why Did They Kill? Cambodia in the Shadow of Genocide* (2005) and nine edited or co-edited collections on violence and genocide.

Gisele Iecker de Almeida is a historian and historical theorist specialising in state-sponsored attempts to overcome difficult pasts. She is a Fellow of the Brazilian Federal Agency CAPES, and is undertaking her doctoral research at the History Department of Ghent University. She has been a visiting researcher at King's College Brazil Institute (2016) and the School of Law and Politics of Cardiff University (2017). Her area of research is the interconnection between the changing nature of our relation with the past and its effects in the present. Ms de Almeida is a member of the steering committee of the International Network for Theory of History and is a founding member of the Latin American Network Historia Pensada.

Victor Igreja is a Senior Lecturer of International Studies at the University of Southern Queensland. His ethnographic and quantitative research focuses on the study of embodied accountability and the political, social and temporal mechanisms to deal with legacies of serious violations in contexts of mass political violence in Mozambique and recently in Timor Leste. The results of his research were published in various social science journals such as *Current Anthropology, Anthropological Quarterly, International Journal of Transitional Justice, Comparative Studies in Society and History, Journal of Traumatic Stress, Review of Political Studies, Nordic Journal of Human Rights, Journal of Southern African Studies, Journal of African Law, Journal of Religion in Africa, Journal of the Royal Anthropological Institute, Journal of Nervous and Mental Disease, British Journal of Psychiatry, Infant Mental Health Journal, Transcultural Psychiatry* and *Encyclopedia of Transitional Justice* and the Wiley-Blackwell's *International Encyclopedia of Anthropology*. Other articles were published in numerous edited volumes.

Natascha Mueller-Hirth is a Lecturer in Sociology at Robert Gordon University in Aberdeen. Her research utilises qualitative, often ethnographic, methodologies to examine issues around peace, conflict, development and gender in South Africa and in Kenya. She has published on NGOs and civil society, the governance of development, racial identities, corporate social responsibility, and time and temporality in transitional societies, and is co-author of *Victim Centred Peacebuilding: Everyday Life Peacebuilding in Northern Ireland, South Africa and Sri Lanka* (2018). Her current research is a study of women's demands for sustainable peace, exploring the experiences and reparative needs of women survivors of the post-election violence in 2007/08 in Kenya.

Sandra Rios Oyola is a Lecturer at the Department of International Studies at Leiden University. She is the author of *Religion, Social Memory and Conflict: The Massacre of Bojayá in Colombia* (Palgrave Macmillan, 2015) based on fieldwork research in the Pacific lowlands of Colombia. As a postdoctoral researcher, she conducted research at the Netherlands Institute of Human Rights, Utrecht University in a large international project funded by the NWO-ORA on 'The Impact of Transitional Justice on Democratic Institution-Building'. She obtained her PhD in Sociology at the University of Aberdeen thanks to a scholarship by the Leverhulme Programme Compromise after Conflict.

Valérie Rosoux is a Senior Research Fellow at the Belgian National Fund for Scientific Research (FNRS). She is a member of the Belgian Royal Academy and teaches *International Negotiation* at the University of Louvain. In 2010–2011, she was a Senior Fellow at the United States Institute of Peace (Washington DC). As a postdoctoral researcher, she worked at The Johns Hopkins University School of Advanced International Studies (SAIS) in 2002, the Center for International Studies and Research (CERI), Sciences-Po Paris (2001) and the University Laval, Canada (2000). Valerie Rosoux has a Licence in Philosophy and a PhD in International Relations.

Eva van Roekel is a Lecturer and Postdoctoral Research Fellow in Cultural Anthropology at Utrecht University. For her PhD she carried out ethnographic research on the relationship between emotion and justice at the trials for crimes against humanity in Argentina from a victims and perpetrators' perspective. Her main research interests include emotion, phenomenology, visual anthropology, medical anthropology, Latin America and military culture. She has published several articles in anthropological journals and has made three documentaries. She is co-founder of Dokumento, a cultural platform for storytelling. Her current research focuses on military mental health in Argentina and UN peace missions.

1 Introduction

Temporal perspectives on transitional and post-conflict societies

Natascha Mueller-Hirth and Sandra Rios Oyola

Introduction

After participating in large international comparative research projects on transitional justice in post-conflict societies, in addition to our own trajectories as researchers specialised on the South African and Colombian cases respectively, we found that reflection on time and temporality, although present in some works, was lacking as a systematic perspective in the fields of transitional justice and peacebuilding. This is surprising, given that time is a crucial dimension in the understanding of social transformation and is particularly relevant for addressing questions related to the legacies of violent pasts and the construction of peaceful futures. The idea for this book drew from Mueller-Hirth's (2017) article on *temporalities of victimhood* and was developed thanks to the generous response by the contributors to the volume. The backbone of the collection is our belief that by exploring how time is experienced, constructed and used by people and institutions, some central problems in post-conflict societies can be revealed and analysed in a unique way. The volume aims, empirically, to develop a deeper understanding of the role time plays in overcoming violent pasts and, theoretically, to contribute time-sensitive perspectives to the fields of transitional justice and peacebuilding.

Our argument is grounded in the sociology of time, which regards measurements of time and notions of temporality as defined by culture and society and examines how time is used to govern and to construct social meanings (Adam 1990; 2004; Bergmann 1992; Elias 1993; Schwartz 1974; Zerubavel 1981, among others). There is no unique ontological manner of registering time nor of perceiving it. In theories of time, a distinction is often made between social time and inner time (see Adam 2004 for an excellent overview). Social time is linear, measurable, predictable, as well as regular and uniform because, as Zerubavel (1981; 1985) claims, social events reflect particular dominant temporal expectations. By contrast, inner time is multiple and discontinuous; it is a cyclical time in which events recur and repeat themselves. Time typifications such as cyclical/circular vs. linear are also debated in the social sciences, particularly in anthropology where cyclical time is often imposed on traditional or small-scale societies and 'Western time' characterised as linear, abstract and sequentially stretched. Munn (1992: 101) has been critical of these ways of typifying time and proposes alternatives such as the image of a leaf, which represents the growth of generations from the originating past and as such involves repetition, successiveness and developmental components. Similarly, Adam (2004: 144–145) suggests that we think of temporal relations as 'clusters of temporal features', such as time frames, tempo, timing, sequence and patterns, whose relationships and relative importance are dynamic and contextually dependent.

While spatial metaphors, such as *moving forward* and *leaving the past behind*, help us to make sense of the *passing of time*, these metaphors are not universal. For example, scientists that rely on geological excavations tend to refer to the past in vertical terms, as that which is deep underground (Simonetti 2013). Understandings of time can follow multiple directions, and different notions of temporality may overlap or clash against each other. Despite the complexity and richness in the social study of time, Western normative and teleological notions of time associated with progress and linearity have predominantly influenced scholars and practitioners in the fields of transitional justice, reconciliation and peacebuilding.

Moreover, to recognise that there are non-linear, and potentially multiple, temporalities does not imply that these temporalities cannot be normative or dominant (Bastian 2014: 15). The social science knowledge on time also allows us to better understand its power effects, demonstrating that control over time is a medium of hierarchical power and governance. There are inequalities in how time is used and whose time is valued, and time plays an important role in social methods of inclusion and exclusion (Bastian 2014). Scholars of time have demonstrated a range of ways in which time becomes the object of social control and temporal power relations are shaped. For example, temporal power relations are gendered; women do more unpaid work and more caring work. Caring work in particular has a temporal logic – fluid, relational and cyclical – that can be contrasted with the linearity of market capitalism and male-dominated productive work (Bryson 2007), while the non-linear and unpredictable 'generativity of child bearing' is sometimes thought of as 'feminine time', or indeed 'feminine timelessness' (West-Pavlov 2012: 101–102). Women are most affected by the need to 'straddle multiple temporalities' (Everingham 2002, cited in Bryson 2007: 134). Representations of singlehood and marriage, and their associated temporal experiences of waiting, moreover, are gendered (Lahad 2012). Temporality is also understood, and new temporal logics shaped, through sexuality and sexual difference (Halberstam 2005). Moreover, temporalities and temporal relations have been shown to operate through unemployment (Auyero 2012, Harms 2013), chronic illness (Charmaz 1991) and drug addiction (Reith 1999), to name but a few areas of research.

Notions of time and temporality play a crucial role in the study of post-conflict societies as well as in the transitional justice paradigm. Transitional justice aims to address wrongdoings of repressive predecessor regimes in order to combat denial and promote justice, accountability and transparency through strengthening the rule of law (Teitel 2003). More recently, transitional justice mechanisms have also been used in established democracies such as Canada or Australia as a form of redress the unfair treatment of indigenous people through reparations and public apologies.

Transitional justice has been described as 'Janus-faced' in the sense that it contributes towards accountability measures that deal with the past as well as mechanisms that seek to assert stable futures.[1] In this regard, Teitel (1997: 2014) states that 'law is caught between the past and the future, between backward-looking and forward-looking, between retrospective and prospective'. Similarly, Hobbs (2016: 513) argues that transitional justice is different from other forms of justice:

> The great significance of the field is its Janus-faced nature. While constitutional redemption is purely forward looking, transitional justice recognises that the past is always with us. In mediating the past, the present and the future, transitional justice understands that political commitment to a more just future can only be secured by acknowledgement of past wrongs and the promise of reparative action.

The terms 'transitional' and 'post-conflict' have clear temporal referents. The promise of transformation is implicitly built upon a notion of progress, 'leaving the past behind' and 'moving toward democracy'. Following an implicit teleological temporality, the transitional justice discourse considers that a society goes from one stage of violence to another of democracy and peace. Transitional justice mechanisms such as trials, truth commissions and amnesties act as seals that control what is brought from the past, such as previously denied human rights abuse, as well as warranting the locking away of the past through amnesties, vetting and trials. They seek to open the doors for future peaceful societies. This teleology is associated with a bias towards Western-style democracies, in which the term 'transition' implies a 'change in a liberalizing direction' (Teitel 2000: 13). Transitional justice's teleological dimension has been criticised for not representing the empirical reality of transitional societies, nor the 'subjective perspective of members of a given transitional society' (Murphy 2017: 68; also see Hinton, Chapter 3 in this volume). Similarly, reconciliation is embedded in a discourse of progress, where societies are moving towards peace by healing broken relationships. In this volume (Chapter 2), Valérie Rosoux explores some of the limitations of this temporal perspective for an understanding of reconciliation.

Moreover, this rhetoric requires a clear setting of boundaries between past and present, as well as future. Amid the chaos of lengthy conflicts, or conflicts that have risen out of chronic deprivation and inequality, setting such boundaries for the past or for the beginning of a conflict in the official discourse is a political act in itself. Definitions of the past are achieved through the implementation of temporal margins for truth commissions, as well as setting temporal limitations for defining who can be considered a victim. In the case of South Africa, but also among many Latin American cases, these temporalities established by the law leave victims of colonialism absent from the official narrative of transition and democratisation, and 'longer temporalities and structural interpretations of the origins of armed conflict disappeared' from the official and public discourse (Castillejo-Cuellar 2014: 51). Indeed, we might argue that, by designating a country a 'post-conflict society', violence is relegated to the past and treated as a temporary episode rather than as an ongoing structural concern.

In the literature on peacebuilding, the concern for time has centred on the *timing* of efforts of resolution, which includes conceptions of 'ripeness': a situation in which 'substantive answers are fruitless until the moment is ripe' (Zartman 2000: 225; also see Rosoux, Chapter 2). This approach focuses on models that can estimate the right moment to advance peace negotiations when other solutions are not achievable. Notions of timing have also been influenced by sequentialism, which underscores the necessity of considering the order of interventions in transitional societies in order to reach the desired effect of peaceful transitions. This perspective warns of 'the dangers of moving quickly toward elections in countries with little democratic history' (Carothers 2007: 15).

An important response to the hegemonic one-size-fits-all model of the liberal peace has been brought by the so-called 'local turn in peacebuilding', which instead emphasises the necessity of empowering local people as the primary authors of peacebuilding (see for example Mac Ginty and Richmond 2013). However, this literature has not explicitly engaged with temporal aspects of peacebuilding, and particularly with the significance of local temporalities that – as several of the contributors to this book demonstrate – shape understandings and practices of violence and peace. Although there has been an interest in issues of timing (Mac Ginty 2016), the lack of attention to specific local understandings of time in the peacebuilding literature seems a missed opportunity.

The time dimensions studied in this book

The contributors to this volume show that a progressive temporality does not necessarily apply to the transitional and post-conflict societies studied and highlight that violence and non-violence are not as easily demarcated. Past and present can seem continuous, rather than being separated clearly as violent/peaceful and democratic, because violence has not ceased or because experiences of marginalisation continue in the 'post'-conflict era. Violence and suffering are not necessarily temporary ruptures but, for many, conflict, violence and suffering are part of the social fabric (Vigh 2008). Moreover, people's lived experiences of time do not always conform to a linear temporality: for instance, victims' traumatic experiences can exist in plural temporalities, answering to the challenge of honouring the past while moving forward.

This collection aims to problematise, through empirical research in a range of contexts, a straightforward, taken-for-granted notion of time in transitional justice and peacebuilding. We have identified several key themes that can be made visible through the lens of time and temporality, which appear and sometimes overlap across the sections and chapters of this volume:

1 tensions, clashes, and negotiations between different temporalities in the context of transitional justice and victims' and perpetrators' experiences of lived time;
2 how these different temporalities produce an unequal distribution of power in post-conflict settings, leading to the socio-temporal marginalisation of some in society and to repercussions for processes of reconciliation, reparations and change;
3 the effects of social acceleration on transitional and post-conflict societies; and
4 how collective memorialisation becomes a vehicle to transmit memories of the past through the lenses of the present.

This introduction now develops these four key themes and then outlines the structure of the book and the contributions that comprise the volume.

Multiple temporalities

A number of contributors found a distinction between temporalities shaped by institutions such as those in charge of reparations, trials and truth commissions, and the temporalities corresponding to the lived experiences of people affected by violence. In the Argentinian context (van Roekel, Chapter 5 in this volume), victims' histories of the last dictatorship and its aftermath were a 'collective jumble of significant personal episodes that did not follow an established chronological thread of events'; conversely, for military officers, by institutionalising and materialising violence, time was an 'orderly and fixed business' with emphasis on the future. In communities in post-war Mozambique (Igreja, Chapter 6), embodied accountability, such as the intrusion of spirits of the dead, pursues liability for serious violations committed over an indeterminate time span and irrespective of the passage of time. These embodied accountability practices can be contrasted with the temporality of global accountability projects, as encapsulated in transitional justice, which involve predetermined time frames and a clear division between past and present.

Importantly, these contributions demonstrate alternative conceptions of time that are used to engage, cope and survive the long-term effects of atrocity and other forms of human rights violations. They highlight that local workings of time shape practices of remembering, reconstruction and accountability. However, although it is crucial for our understanding of societies emerging from conflict and authoritarianism to recognise the existence of these different temporalities, this recognition can have disempowering effects for marginalised people, given the hierarchies and power differentials in the uses and understandings of time.

The idea of the disempowering effect of the notion that certain groups belong to different temporalities is not new. Johannes Fabian (1983: 1), in his analysis of the uses of time in anthropology, exposed that certain time discourses are employed in the description of the 'other' and function to create distance in a seemingly neutral manner or in describing certain groups as archaic, tribal or primitive. The 'temporalising of difference' can involve the labelling of 'other' people, cultures and societies as belonging to the past, which gives rise to associated claims about their irrationality, underdevelopment, rituals, religiosity, and so on (Helliwell and Hindess 2005).

Time and unequal power relations

We can observe this link between time and power in relation to victims/survivors in post-conflict societies, who, after a certain amount of time has passed, are often portrayed as anachronistic, as belonging to a different time, being unwilling to 'move on' into the democratic era, or failing to conform to the dominant post-conflict temporality. Victims/ survivors, but also other groups in society, can feel 'lost in transition' (Pedersen and Højer 2008) or 'out of sync' with the rest of society (Mueller-Hirth 2017: 200). Societal pressure to forget the past or to forgive can result in a denial of their rhythm of possible reconciliation in favour of political reconciliation demanded from governments facing vulnerable transitions, as is investigated by Rios Oyola in relation to Colombia and by Benda with regards to Rwanda in this volume (Chapters 4 and 8, respectively). Benda explores the different versions of future and past used by policy makers in post-genocide Rwanda. Rios Oyola investigates how the temporality of political peace processes influences the temporality of social peace processes, analysing this relationship in terms of social acceleration. Social and political expectations about the timing, duration and pace of people's victimhood and their willingness to forgive, which reflect linear 'transitional justice time' (Hinton, this volume) and its notions of the past, present and future, lead to the socio-temporal marginalisation of people's lived experiences.

An exercise of power through the politics of time can also be observed in the delays and waiting times that people experience in post-conflict societies, for example in relation to reparations, victim support, recognition or socio-economic change. The distribution of waiting time is connected with the distribution of power: to be kept waiting is a social assertion that one's time and social worth are less valuable (Schwartz 1974; Bourdieu 2000). Because waiting is bound up with social status and power, it is often designated to the poor and the powerless (Harms 2013; Auyero 2012), and indeed exacerbates inequalities (Reid 2013). Waiting, as analysed by Mueller-Hirth in Chapter 7 of this volume, is a time-based problem that reveals the complexities and inequalities of reparations and change in South Africa more than two decades after the end of Apartheid.

The effects of acceleration

Part of the inequalities that the authors observe stem from a temporal conflict between social acceleration and the specific demands of transitional and post-conflict societies. Social acceleration has characterised the development of modernity, although the emergence of a technologically-driven neoliberalism and the increasing networking of societies has more recently led to increasing time scarcity and the 'shrinking of the present' (Rosa 2003; Rosa and Scheuerman 2009). Rosa (2003) distinguishes three categories of acceleration: technological acceleration, the acceleration of social change and the acceleration of the pace of life. We contend that acceleration throws up particular issues for transitional and post-conflict societies.

In general terms, the pace of economic and social change in 'acceleration societies' (Rosa 2003) involves a very different temporal logic than the logic underlying victim support and reparations policies, which are often aimed at redressing the legacies of many years of violence and discrimination. Virilio (1997) has claimed that we live in a 'dictatorship of speed'; the necessity of speed is held up as a virtue and is associated with vitality and vigour (Tomlinson 2007). Conversely, patience, something that is often asked of people in transitional and post-conflict societies, is a cultural value that arguably loses its significance with the development of modernity and the compression of time. We contend here that there are conflicts between the demands of acceleration societies and the patience that is required in post-conflict societies in order to achieve redistribution, redress and social reconstruction. Acceleration, and its associated virtue of speed, provides a temporal frame for social expectations on victims to move on and for transitional justice processes to wrap up quickly. For example, conflicts between different time cultures have been documented in Germany after reunification, when West German depictions of East German culture portrayed it as too slow and always being 'behind in time' (Rau 2002).

Moreover, acceleration is relevant to the concerns of this volume in that political apologies and other political processes can act as accelerators for reconciliation efforts, without important historical, social and economic issues necessarily having been addressed (also see Rosoux's chapter on the tempo of reconciliation, this volume). Political apologies can be understood as 'prime examples of deliberate attempts to intervene in shared understandings of political community and its temporality' (Bastian 2013: 3). Rios Oyola's chapter (this volume) seeks to understand to what extent the Colombian peace plebiscite and the FARC's public apologies work as fuel for accelerating the social peace process in order to overcome desynchronisation with the political peace process. The challenges emerging from this desynchronisation echo the pressures on victims in relation to forgiveness, as already outlined above (2).

Memorialisation

Transitional justice discourse holds the promise of remembering as a way to 'never again' perpetrate the types of atrocities that occurred in the past.

Memories, as acts of representations, allow individuals to locate themselves in time and distinguish themselves from the past (Feindt et al. 2014: 28). The impact of memory in transitional and post-conflict societies is found in the framework of a *politics of regret*; the politics adopted by post-conflict countries through attempts to retrospective justice and redressing past wrongs (Olick 2007). The foundation of the idea that collective memory offers a strong relationship between the past and the present relies on the notion that societies are capable of learning lessons from the past, and that proper narratives of social memory can help to prevent societies from romanticising past violence. Politics of memory are intended to work as careful warnings against the temptation of nostalgia by recognising the unjust harms suffered by victims but they also 'propagate new interpretative narratives about the "what happened" to legitimate a new political dispensation and develop a new vision of the future for the polity' (de Brito 2010: 360, also see Iecker de Almeida, Chapter 9).

Just as there is no present without past there is also no past without present, as Iecker de Almeida demonstrates when she writes about acts of *undoing the past* in post-authoritarian Brazil: 'undoing is the attempt to make it seem *as if* the past had never happened by creating different timelines which allow the adoption of an alternative future' (Chapter 9, this volume). Both official policies of memorialisation and spontaneous grassroots memorialisation are influenced by present events and present emotions but 'the past is [also] re-evaluated in light of the future' (Assman 2013: 55). Current collective emotions act as gateways of memories of the past; they help to create representations of the past according to the interests of the present, following Koselleck's claim that 'the present past is saturated with personal memories and emotions through which the living are involved in it and bound up in it' (Assman 2013: 48). It is possible that current social emotions influenced by a human rights discourse could help to shape narratives of social memory that allow understanding the past as lessons for the future and preventing the repetition of violence.

The mere recollection of the past is not sufficient warrant for such events to not be repeated. The power of remembering in order for such kind of atrocities to not happen 'never again!' should be based on a clear peacebuilding agenda, since today's victims, justified by their memories, can become tomorrow's perpetrators. Mamdani (2002: 17) illustrates this point in the case of the Rwandan genocide: 'How many perpetrators were victims of yesteryears? What happens when yesterday's victims act out of a determination that they must never again be victimised, *never again*?' Volkan's (1997; 2004) notion of chosen traumas – shared mental representations of a massive trauma experienced by a group's ancestors that are reactivated and sometimes exaggerated – also emphasises how past violations can drive present-day social movements and political mobilisations. Memory representations can lead to a 'time collapse', where ideas and feelings connected with the past are folded in with those of a current conflict: 'under the influence of a time collapse, people may intellectually separate the past event from the present one, but emotionally the two events are merged' (Volkan 1997: 35).

It is important to note however that, just as there is no linear narrative of violence, there is no linear teleological transmission of memories of violence (Shaw 2010). While some memories of violence can cohere into approved forms, other forms of memory are embodied, ritual, non- or poly-discursive (Berliner 2010; Pichler 2010; also see Igreja on embodied accountability in this volume). Rothberg (2009: 13–14) introduces the concept of multi-directional memory in order to highlight how memories are subject to competition, negotiation and overlapping led by social and political forces. Gandolfo, in Chapter 10, considers ruins in Palestine/Israel as examples of embodied memory and resistance. She explores how walking through ruins represents the visitor's negotiation with 'pluritemporalities' – different interpretations of collective and subjective time. Gandolfo shows the ruins she studied as an example of sites that allow memory practices to be revisited, redefined and relocated through ongoing processes of memory work, such as cultural activism and heritage movements. The notion of a clear division between past and present then needs to be challenged:

> Many of the salient phenomena in international and domestic politics of the last decades – reparation politics, the outing of official apologies, the creation of truth commissions, historical commissions and commissions of historical reconciliation, etc. – revolve around a growing conviction that the once commonsensical idea of a past automatically distancing itself from the present is fundamentally problematic, and that the belief that the past is superseded by every new present has been more a wish than an experiential reality.
>
> (Lorenz and Bevernage 2013: 16)

From this perspective, past, present and future are less part of a linear continuity but overlap, revolve in circularities and can influence each other. The challenge to describe, define and understand these multiple temporalities is one that we have encountered in this book.

The relevance of this volume

We agree with Barbara Adam who writes that we need 'time literate social theory' (2003). In this vein, we want to advance time-literate thinking about the issues that affect transitional and post-conflict societies. The contributors to this book all seek to develop a deeper understanding of how time is experienced and constructed in transitional and post-conflict societies through different uses and understandings of temporal categories and concepts. Research in the areas of transitional justice and peacebuilding has rarely systematically considered these issues.

However, as Munn (1992: 116) noted, the problem of time is the difficulty of 'find[ing] a meta-language to conceptualise something so ordinary and apparently transparent in everyday life.' How can we take time more explicitly into account when we study transitions and post-conflict societies, and how can temporal analyses provide insights into their social and political dynamics? In this regard, this volume links social theories of time to empirical research from a wide geographical range of case studies. The different conceptions of time and temporality used in this volume do not attempt to provide a unified theory of time in transitional societies; rather, the authors present temporal analyses of some of the key dynamics of transitional and post-conflict societies, such as victimhood, reconciliation, forgiveness, collective memory building, intergenerational transmission and reparations.

Therefore, while this book develops a novel theoretical approach to studying transitional and post-conflict societies, its key objective is to enhance our understanding of the challenges these societies face after the ending of mass violence. To engage with dimensions of time and temporality allows researchers, policy makers and practitioners to problematise the discourse, implementation and long-term consequences of transitional justice in ways that have not been observed before. Different perceptions of time and temporality have many significant impacts for the operationalisation of policies aimed to transform the aftermath of conflict. For example, institutions that rely on the reporting of instances of violence and conflict in order to develop future-oriented policies rely on the accuracy of reports that are based on Western notions of time that often do not reflect local understandings of conflict and temporality (Igreja 2012; McLeod 2013) or the consequences of living amid chronic and acute violence (Read and Mac Ginty 2017). The 'politics of waiting', as they play out in relation to victim support policies, can shape people's engagement with the state and former enemies in the post-conflict settlement.

The research in this volume demonstrates that 'transitional justice time' is not neutral and that temporal modes produce particular power relations in transitional and post-conflict settings. A notion of linear time, if left unchallenged, might be divisive in narratives about societies 'moving forward' without recognising the multiple temporalities that exist in relation to traumatic or violent experiences and that can account for how people's conceptions of justice and healing change dynamically (Igreja 2012). Moreover, we believe that developing a deeper understanding of the role time plays in overcoming violent pasts and the challenges, many of them long-term, that are faced in post-conflict societies can bring us closer to the lived reality of peacebuilding and transitional justice and to the people who are most affected by them. Conflicts are often triggered by the 'historical accumulation of separate and unresolved violent conflicts' (Igreja 2012: 4) or 'chosen traumas' (Volkan 2004), as discussed above, that can lead to renewed cycles of violence. These cycles are in no manner unescapable, and an appropriate understanding of how time is perceived, constructed and lived in transitional and post-conflict societies can help to address and prevent ritualised violence and the repetition or re-enactment of traumatic memories. We hope that this new understanding goes some way towards better grasping the conditions for establishing sustainable peace.

As such, our objectives in this volume are both practical and theoretical: this book not only intends to make a case for the need to theoretically engage with the category of time, but we also hope that, in its empirical contributions, it can highlight some neglected dimensions that are crucial for building sustainable peace. Similarly, our focus on temporalities and temporal domination does not imply that we neglect other dimensions of power, marginalisation and exclusion. On the contrary, one of our concerns in this volume is to pay closer attention to taken-for-granted aspects of transitions and less noticeable forms of power, and to become alert to how socio-temporal marginalisation contributes to the exercise of power.

Structure of the book

This book presents original research from nine transitional or post-conflict settings, including Argentina, Brazil, Colombia, Chile, Rwanda, Cambodia, Mozambique, Palestine/Israel and South Africa. The authors have used different approaches, methodologies and conceptualisations in analysing aspects of time and temporality. Their research is based on interviews, participant observations, document analysis and historical analysis, among other methods, and is often ethnographic in approach. Together, the authors demonstrate that temporal analyses can provide new and important insights into the social and political dynamics of societies after the end of violence. Although the collection provides an ambitious geographical scope, its purpose is to investigate and problematise the role of time and temporality in post-conflict and transitional societies, rather than providing exhaustive description of cases or cross-comparative analysis.

The volume is divided into three sections. The sections have emerged from, and reflect, the empirical cases the contributors produced. The four substantial themes that we developed earlier – multiple temporalities, time and power, acceleration and memorialisation – cut across these sections. Part I, *Questioning Transitional Justice Time*, establishes that transitions, transitional justice institutions and processes have particular dominant time orientations that are associated with truth claims. In Chapter 2, Valérie Rosoux provides an opening conceptual reflection on the necessity of considering time in post-conflict negotiations and reconciliation, introducing many of the issues that subsequent chapters unpack, such as timing, tempo and closure. She criticises the idea of the 'ripeness' of a conflict by stimulating a comparative analysis between different cases, including the Franco-German, Franco-Algerian, Rwandan and South African conflicts. The chapter demonstrates that differences in tempo and timing challenge enthusiastic, but premature, calls for reconciliation. Reconciliation emerges as an open-ended and non-linear process that is shaped by a plurality of interests and a multiplicity of time frames. Chapter 3, by Alexander Hinton, draws on ethnographic research on the Extraordinary Chambers in the Courts of Cambodia. This so-called Khmer Rouge Tribunal, a hybrid tribunal with an outreach mandate, is an example for growing awareness of the gap between abstract justice and the understandings and experiences of post-conflict populations. Through analysing the journey of Uncle San and Aunty Yan, characters in a widely used Tribunal outreach graphic booklet, Hinton argues that a particular conception of time – 'transitional justice time' – is manifest in its narrative. Transitional justice time is premised upon a particular conception of teleological transformation and idealised liberal democratic being that serves to erase historical complexities and masks, and even effaces, local understandings, including Buddhist temporalities prevalent in Cambodia. In Chapter 4, Rios Oyola analyses how the urgency of the political peace process in Colombia is subjected to deadlines that reflect the vulnerability of the agreement, political interests and the pressure of the international community. This urgency is transmitted to the social peace process, understood as the process of societal healing conducted by civil society (Brewer 2013). Her chapter does not only highlight the existence of different temporalities between the social and the political peace process, but studies how the latter tries to influence the former through social acceleration devices such as the peace plebiscite and public apologies.

This second part of the book, *Co-existing and conflicting temporalities*, attends to victims' and perpetrators' experiences of lived time, which are often at odds with those of 'transitional justice time' as examined in Part I. Eva van Roekel studies, from an anthropological perspective, the relationships between time and violence in Argentina in Chapter 5. Her data, emerging from conversations with victims and perpetrators, illustrate how different and dynamic forms of time-as-lived and valuations of violence co-exist and produce diverse understandings of local transitional justice trajectories. She says that we should rethink transitional justice processes by analysing the specific nature of how communities and individuals conceive the relationship between violence and time: far from being linear movements away from violence and suffering, transitional justice processes can exist as an ongoing re-enactment of the violence and suffering and a moral practice to keep it in the present. Victor Igreja, in Chapter 6, similarly addresses the contrast between the temporality of local practices of accountability and of forms of global accountability that are part of the transitional justice repertoire. Drawing on ethnographic research and the case of a family in a community in conflict in Mozambique, he shows that liability and justice remain intrinsically linked to embodied processes that transcend time and individual selves. Consequently, addressing injustice does not necessarily involve conflict resolution such as it would be upheld by institutions, but rather an embodied accountability across generations. His research suggests that negotiations over temporalities of wrongdoing, instead of stricter periodisation of gross human rights violations, are instrumental for averting new cycles of vengeful violence. Chapter 7 moves from questions of accountability to concerns about the timing of reparations and recognition. Mueller-Hirth investigates the experiences and effects of waiting for victim support and social change among those affected by Apartheid-era human rights violations, twenty years after the democratic transition. She employs the notion of 'fixed liminality' to argue that the need to wait, and the uncertainties associated with it, has left some victims/ survivors in a permanent in-between state and unable to move on from their victimhood. Her contribution demonstrates that delays and waiting are strategies of power that contribute to the socio-temporal marginalisation of particular groups in post-conflict societies, making the timing of reparations a crucial issue.

The third part of the book, *Intergenerational transmission and Memorialisation*, is concerned with the relationships of time with memory in relation to how pasts, presents and futures are understood and mobilised. Richard Benda examines important issues around the intergenerational transmission of trauma, guilt and responsibility in his chapter on the 'children of killers' in post-genocide Rwanda. Drawing on participatory research with Youth Connekt Dialogue (YCD) – essentially *intragenerational* dialogues that deal with the *intergenerational* transmission of trauma and guilt – he shows that the government's claims to be the sole custodian of the social organisation of time is contested both by victims and by children of perpetrators. Through the metaphors of 'rear-view mirror' versus 'windshield' temporalities, shared intergenerational time senses and a subaltern temporality could be expressed. He demonstrates that, by applying a time-focused approach to YCD, we can gain a better understanding of how plural, competing and complementary temporalities interact in the transitional timeline, to reorder the past and shape the future, and critique the government's temporal expectations for national reconstruction and reconciliation. In Chapter 9, Gisele Iecker de Almeida critically analyses the philosophy of history that underlies the processes of re-memorialisation of Brazil's most recent dictatorial period. By analysing the reparation and truth seeking programmes and recent state efforts at memorialisation and renaming, she traces the ways in which the Brazilian state has sought to 'undo the past' by reshaping how it is perceived in the present in line with a more desirable future and new narrative of the nation. Moving on from the previous chapter's concern with a state's efforts to shape memory to a study of restoration as a form of activism, Luisa Gandolfo, in Chapter 10, considers the depopulated West Bank towns of Kufr Bir'im and Iqrit. She explores the ways in which restoration of these sites provides a point of memorialisation, healing and communal commemoration, as well as a mechanism of resistance and cultural activivism. In doing so, she counters Halbwachs' idea that history is a 'dead memory' by positing that an 'organic' relation is sustained through the presence of subsequent generations who keep history alive through formal and informal commemorations and celebrations. By engaging with the notion of pluritemporality, her chapter considers the position of the sites as material mnemonics, physically present, yet absent from the state discourse. As Palestinians and Israelis ascribe alternative names and histories to the sites, the locations become not just heterotopic counter-sites, but also heterochronic as certain epochs are selected and others discarded in the retelling of a site's past.

Finally, the conclusion briefly summarises the main arguments of the preceding chapters. A key shared finding across the diverse cases concerns the differences, and frequent temporal conflicts, between local, or lived, temporalities and institutional, or official, temporalities. The chapter reflects on the implications of this finding and on the relevance – both theoretical and in terms of policy – of the collection's focus on time and temporality in transitional justice and peacebuilding.

Note

1 Crawford (2015: 475) presents a similar claim for restorative justice 'simultaneously look[ing] backwards and forward across time'.

References

Adam, Barbara. 1990. *Time and Social Theory.* Cambridge: Polity.

Adam, Barbara. 2003. Reflexive modernization temporalized. *Theory, Culture & Society* 20(2): 59–78.

Adam, Barbara. 2004. *Time.* Cambridge: Polity.

Assman, Aleida. 2013. Transformations of the Modern Time Regime. In Chris Lorenz and Berber Bevernage (eds), *Breaking up Time: Negotiating the Borders between Present, Past and Future*, 39–56. Göttingen: Vandenhoeck & Ruprecht.

Auyero, Javier. 2012. *Patients of the State: The Politics of Waiting in Argentina.* Durham, NC: Duke University Press.

Barahona de Brito, Alexandra. 2010. Transitional justice and memory: Exploring perspectives. *South European Society and Politics* 15(3): 359–376.

Bastian, Michelle. 2013. Political apologies and the question of a 'shared time' in the Australian context. *Theory, Culture & Society* 30(5): 94–121.

Bastian, Michelle. 2014. Time and community: A scoping study. *Time & Society* 23(2): 137–166.

Bergmann, Werner. 1992. The problem of time in sociology: An overview of the literature on the state of theory and research on the sociology of time, 1900–1982. *Time & Society* 1(1): 81–134.

Berliner, David. 2010. Memories of Initiation Violence: Remembered Pain and Religious Transmission among the Bulongic (Guinea, Conakry). In Nicolas Argenti and Katherina Schramm (eds), *Remembering Violence: Anthropological Perspectives on Intergenerational Transmission*, 39–56. Oxford: Berghahn Books.

Bourdieu, Pierre. 2000. *Pascalian Meditations.* Redwood City, CA: Stanford University Press.

Brewer, John. 2013. Sociology and Peacebuilding. In Roger Mac Ginty (ed.), *Routledge Handbook of Peacebuilding*, 159–170. London: Routledge.

Bryson, Valerie. 2007. *Gender and the Politics of Time: Feminist Theory and Contemporary Debates.* Bristol: Policy Press.

Carothers, Thomas. 2007. The sequencing fallacy. *Journal of Democracy* 18(1): 12–27.

Charmaz, Kathy. 1991. *Good Days, Bad Days: The Self in Chronic Illness and Time.* New Brunswick, NJ: Rutgers University Press.

Castillejo-Cuellar, Alejandro. 2014. Historical injuries, temporality and the law: Articulations of a violent past in two transitional scenarios. *Law Critique* 25: 47–66.

Crawford, Adam. 2015. Temporality in restorative justice: On time, timing and time-consciousness. *Theoretical Criminology: An International Journal* 19(4): 470–490.

Elias, Norbert. 1993. *Time: An Essay.* Oxford: Blackwell.

Fabian, Johannes. 1983. *Time and the Other: How Anthropology Makes its Object.* New York: Columbia University Press.

Feindt, G., Krawatzek, F., Mehler, D., Pestel, F. and Trimçev, R. 2014. Entangled memory: Toward a third wave in memory studies. *History and Theory* 53(1): 24–44.

Halberstam, Judith. 2005. *In a Queer Time and place: Transgender Bodies, Subcultural Lives.* London: NYU Press.

Harms, Erik. 2013. Eviction time in the new Saigon: Temporalities of displacement in the rubble of development. *Cultural Anthropology* 28(2): 344–368.

Helliwell, Christine and Hindess, Barry. 2005. The temporalizing of difference. *Ethnicities* 5(3), 414–418.

Hobbs, Harry. 2016. Locating the logic of transitional justice in liberal democracies: Native title in Australia. *UNSW Law Journal* 39(2): 512–552.

Igreja, Victor. 2012. Multiple temporalities in indigenous justice and healing practices in Mozambique. *The International Journal of Transitional Justice* 6(3): 1–19.

Lahad, Kinneret. 2012. Singlehood, waiting, and the sociology of time. *Sociological Forum* 27 (1): 163–186.

Mac Ginty, Roger. 2016. Political versus Sociological Time: The Fraught World of Timelines and Deadlines. In Arnim Langer and Graham K. Brown (eds), *Building Sustainable Peace: Timing and Sequencing of Post-Conflict Reconstruction and Peacebuilding*, 15–31. Oxford: Oxford University Press.

Mac Ginty, Roger and Richmond, Oliver P. 2013. The local turn in peace building: A critical agenda for peace. *Third World Quarterly* 34(5): 763–778.

McLeod, Laura. 2013. Back to the future: Temporality and gender security narratives in Serbia. *Security Dialogue* 44(2), 165–181.

Mamdani, Mahmood. 2002. Making sense of political violence in postcolonial Africa. *Identity, Culture and Politics* 3(2): 1–24.

Mueller-Hirth, Natascha. 2017. Temporalities of victimhood: Time in the study of postconflict societies. *Sociological Forum* 32(1): 186–206.

Munn, Nancy D. 1992. The cultural anthropology of time: A critical essay. *Annual Review of Anthropology* 21(1): 93–123.

Murphy, Colleen. 2017. *The Conceptual Foundations of Transitional Justice.* Cambridge: Cambridge University Press.

Olick, Jeffrey K. 2007. *The Politics of Regret: On Collective Memory and Historical Responsibility.* Abingdon: Routledge.

Pedersen, Morten Axel and Højer, Lars. 2008. Lost in transition: Fuzzy property and leaky selves in Ulaanbaatar. *Ethnos* 73(1), 73–96.

Pichler, Adelheid. 2010. Memories of Slavery: Narrating History in Ritual. In Nicolas Argenti and Katherina Schramm (eds), *Remembering Violence: Anthropological Perspectives on Intergenerational Transmission*, 135–163. Oxford: Berghahn Books.

Rau, H. (2002). Time divided – time united?: Temporal aspects of German unification. *Time & Society* 11(2–3): 271–294.

Read, Roisin and Mac Ginty, Roger. 2017. The temporal dimension in accounts of violent conflict: A case study from Darfur. *Journal of Intervention and Statebuilding*, 1–19.

Reid, Megan. 2013. Social Policy, 'Deservingness,' and sociotemporal marginalization: Katrina survivors and FEMA. *Sociological Forum* 28(4): 742–763.

Reith, Gerda. 1999. In search of lost time: Recall, projection and the phenomenology of addiction. *Time & Society* 8(1): 99–117.

Rios Oyola, Sandra Milena. 2015. *Religion, Social Memory, and Conflict: The Massacre of Bojayá in Colombia*. Palgrave Studies in Compromise after Conflict. Basingstoke: Palgrave Macmillan.

Rosa, Hartmut. 2003. Social acceleration: Ethical and political consequences of a desynchronized high-speed society. *Constellations* 10(1): 3–33.

Rosa, Hartmut and Scheuerman, William E. 2009. Introduction. In Hartmut Rosa and William E. Scheuerman (eds), *High-Speed Society: Social Acceleration, Power and Modernity*, 1–29. University Park: Pennsylvania State University Press.

Rothberg, Michael. 2009. *Multidirectional Memory: Remembering the Holocaust in the Age of Decolonization*. Stanford, CA: Stanford University Press.

Schwartz, Barry. 1974. Waiting, exchange, and power: The distribution of time in social systems. *American Journal of Sociology* 79(4): 841–870.

Shaw, Rosalind. 2010. Afterword: Violence and the generation of memory. In Nicolas Argenti and Katherina Schramm (eds), *Remembering Violence: Anthropological Perspectives on Intergenerational Transmission*, 251–260. Oxford: Berghahn Books.

Simonetti, C. 2013. With the past under your feet: On the development of time concepts in archaeology. *Anuário Antropológico* 39(2): 283–313.

Teitel, R. 1997. Transitional jurisprudence: The role of law in political transformation. *Yale Law Journal* 106: 2010–2080.

Teitel, R. 2000. *Transitional Justice*. Oxford: Oxford University Press.

Teitel, R. 2003. Transitional justice genealogy. *Harvard Human Rights Journal* 16(Spring): 69–94.

Tomlinson, John. 2007. *The Culture of Speed: The Coming of Immediacy*. London: SAGE.

Vigh, H. (2008). Crisis and chronicity: Anthropological perspectives on continuous conflict and decline. *Ethnos* 73(1): 5–24.

Virilio, Paul. 1997. *Open Sky*. London: Verso

Volkan, Vamik. 1997. *Bloodlines: From Ethnic Pride to Ethnic Terrorism*. New York: Farrar, Straus & Giroux.

Volkan, Vamik. 2004. 'Chosen trauma, the political ideology of entitlement and violence'. Paper presented at Berlin Meeting, 10 June. Retrieved 2 September 2014 (http://vamikvolkan.com/Chosen-Trauma,-the-Political-Ideology-of-Entitlement-and-Violence.php).

West-Pavlov, Russell. 2012. *Temporalities*. Abingdon: Routledge.

Zartman, I. William. 2000. Ripeness: The Hurting Stalemate and Beyond. In Paul Stern and Daniel Druckman (eds), *International Conflict Resolution after the Cold War*. Washington, DC: National Academy Press.

Zerubavel, Eviatar. 1981. *Hidden Rhythms: Schedules and Calendars in Social Life*. Chicago, IL: University of Chicago Press.

Zerubavel, Eviatar. 1985. *The Seven Day Circle*. New York: Free Press / Macmillan.

2 Time and reconciliation

Negotiating with ghosts

Valérie Rosoux

> Les morts sont les invisibles, mais ils ne sont pas les absents.
>
> Victor Hugo[1]

Introduction

In the aftermath of international or civil wars, time is pivotal, but in a paradoxical way, urgency matters as much as duration. On the one hand, priorities such as rebuilding the political machinery and the economy, or prosecuting human rights abusers, can be argued as matters of great urgency: '[t]he risk of relapse into violent conflict only increases with time' (UN Security Council S 2011/634: 4). This urgency requires extremely quick reactions on the ground and a sharp sense of adaptability. On the other hand, timing is also of consequence in terms of duration. Speed, as such, is no guarantee of success in the long run. The hopes and disillusions resulting from the Arab Spring – where political changes succeeded one another at a dazzling pace – show that haste can have violent and highly questionable consequences. Thus, the rush to achieve quick results with respect to conflict resolution, and, even more so, reconciliation, may lack strong roots and may be unsustainable. Case studies such as Iraq, Afghanistan, or Bosnia-Herzegovina remind us that time is necessary in order to recognize the scale of destruction of both the infrastructure and the social fabric of the country. If the initial process does not allow this indispensable time, the rest may be built on sand (Barsalou 2005; Sissons and Al-Saiedi 2013).

To consider *both* urgency and duration, scholars and practitioners underline the need to nest short-term stabilization imperatives in the longer-term goal of reconciliation (USIP 2009: 31). However, this focus on a conscious nesting does not prevent ongoing tension between the insistence on quick, observable and high-impact results and the more open-ended aspect of any reconciliation process. In addressing this issue, this chapter is divided into three main parts. The first questions the notion of *timing*. It focuses on the starting point of any reconciliation process. When are people ready to favour a rapprochement towards their former enemies? Which turning point can explain the passage from rivalry to partnership (Schaap 2007)? The second part examines the notion of *tempo*. It emphasizes the long-term character of any rapprochement. After mass atrocities, changes of attitude, emotions and beliefs take time. What is the appropriate pace? The third part puts the emphasis on the notion of *closure*. Until when do the former adversaries refer to the conflictual past? When do they definitely turn the page? Is closure actually possible in all cases?

These sections show that three main factors contribute to the rapprochement between former enemies. First, the contextual variable: in most cases, national and international circumstances dictate the timing (part 1). Second, the personal variable: legitimate leadership – or the lack of legitimate leadership – highly impacts the pace of the rapprochement (part 2). Third, the nature of the past violence: the survivors' experience in various cases demonstrates that extreme violence produces a 'counter-time' that impedes their normal progress through 'ordinary' time (Langer 1991).

This chapter emerges from a larger research project based on a case-oriented and inductive research design. It focuses on four configurations where official representatives commonly refer to the notion of reconciliation: the Franco-German case (post-international conflict), the South African case (post-internal conflict), the Franco-Algerian case (post-colonial conflict) and the Rwandan case (post-genocide).[2] The research stance adopted throughout this chapter is not intended to be cynical. It does not consider *a priori* that the call for reconciliation can be reduced to a slogan. It is not meant to be euphoric either. Reconciliation cannot be a miracle formula to cure all post-conflict challenges.

To use a visual metaphor, the city of Koblenz in Germany illustrates very well what is at stake in this study. It is fascinating to pay attention to the colours of waters at the confluence of the Rhine and the Moselle. Far from mixing immediately, the two rivers keep their respective colours for a rather long time. It is downstream, far beyond the confluence, that the waters of both rivers become progressively integrated and eventually do not differentiate themselves from each other. Like these streams, and this is the main hypothesis of this study, communities that were affected by a violent past can only move closer over an extended period of time – when they can.

Timing: when does it start?

In his powerful pamphlet, *Common Sense* (1776), Thomas Paine advocates colonial America's independence from the kingdom of Great Britain. His argument is sharp: 'Reconciliation *now* is a dangerous doctrine.' In referring to the 'wounds of deadly hate', he argues that any connection with Great Britain would be 'forced' and 'unnatural'. His description of the 'fatal nineteenth of April, 1775' (massacre at Lexington) underlines the temporal limitations of any reconciliation work. His point is categorical: 'The last cord now is broken; the people of England are presenting addresses against us. There are injuries which nature cannot forgive; she would cease to be nature if she did'

(Paine 2008: 82). Identifying with 'the ruined and insulted sufferers in all and every part of the continent', he severely condemns 'those who espouse the doctrine of reconciliation': 'it is easy getting into holes and corners and talking of reconciliation. But do such men seriously consider, how difficult the task is (…)? Do they put themselves in the place of the sufferer whose all is already gone?' (Paine 2008: 122). The interrogation is – and remains, even 240 years later – critical. It reminds us that reconciliation cannot be started at any time.

A deceptive impatience

In this regard, there is a wide gap between external calls for reconciliation and most people's concerns in the aftermath of a war. The vast majority of outsiders highlight the need for reconciliation immediately after the end of the hostilities or even *before* it. A single anecdote illustrates this attitude. During a conference, the American diplomat Melanne Verveer explained that the United States had just launched a vast programme devoted to 'reconciliation projects'. Questioned about what the concept of reconciliation meant to her, the close adviser of Hilary Clinton (who was then Secretary of State) hesitated and then asserted: 'Reconciliation has to happen everywhere, as soon as possible, and at all levels of society' (Washington, 4 November 2010). There was no real answer to the initial definition question. Reconciliation is reconciliation. It has to happen, and more importantly, it has to happen as soon as possible.

The reaction of the former French President, Nicolas Sarkozy, after the death of Colonel Gaddafi confirms this perspective. Just one day after the death of the former Libyan president, Nicolas Sarkozy had a simple message: 'What Libya needs now is reconciliation. Libyans need to forgive each other' (21 October 2011, AFP). Even though the country was still devastated by extreme violence, and despite the ambiguity of the notion of forgiveness in the aftermath of a dictatorship, the reference to reconciliation was almost instantaneous. The same comment can be made more broadly about the attitude of European Union (EU) representatives. As the former European Commissioner for Enlargement and European Neighbourhood Policy, Stefan Füle, put it, 'there is no alternative to reconciliation' (Batumi, 11 July 2012). Associated with notions of peace, stability and reconstruction, reconciliation is presented as necessary – from the Maldives to Afghanistan, via Libya, Turkey, Ivory Coast, Somalia, Soudan and Mali.

This argument has been particularly tangible in former Yugoslavia. Thus, the European Commission recommended starting accession talks immediately after the conclusion of an agreement between Serbia and its province Kosovo. In this respect, reconciliation almost served as an accession condition for joining the EU. At first glance, this persistent pressure was not entirely futile. One of the indications of this is the apparent transformation of the Serbian president's behaviour. A fervent nationalist, Tomislav Nikolic used to refuse even the idea of recognizing Kosovo after his inauguration in June 2012. Less than a year later, however, the economic crisis and the perspective of becoming a member of the EU led him to soften his position towards Pristina. There has been an equally clear change in his attitude towards the crimes committed at Srebrenica. When declaring that he was 'kneeling down' and asking that 'Serbia be forgiven for the crimes committed in Srebrenica' (Belgrade, 26 April 2013), Tomislav Nikolic seemed to comply with the EU norms. However, as this chapter suggests, it takes more than words – essentially words geared towards pleasing Brussels – to open an effective path to reconciliation up.

The same pattern could be detected regarding the United Nations (UN). In Kosovo, reconciliation between Serbs and Albanians was also an explicit aim of the UN peace-consolidation mission. Nevertheless, the impossibility of reaching an agreement as to the status of Kosovo showed that such a purpose cannot be externally imposed. While many claim that a rapprochement between former enemies depends on tangible assistance from the international community, the forces for change are primarily internal and cannot be coerced.

Immediately after the end of hostilities, the population affected by the violence never calls for a rapprochement with those who are still perceived as enemies. The discrepancy between outsiders' optimism and people's actual needs is confirmed by a survey of victims' expectations in Uganda. After two decades of civil war between the Lord's Resistance Army (LRA) and government forces, the people of Northern Uganda were subjected to countless brutalities. Tens of thousands of civilians were killed and/or mutilated; at least 20,000 children were abducted, more than 1.9 million people were displaced from their homes into camps (HRW 2012). A survey carried out in 2005 sought answers to the following question: what should be done for the victims of violence in Northern Uganda? Most respondents said that victims should first be provided with some form of assistance, with 52% indicating financial compensation, 40% mentioning food and 26% education for children. Interestingly, justice is ranked low in the poll in terms of immediate needs (8%), and reconciliation even lower (6%) (ICTJ 2005: 7).

Thus, despite the impatient calls for reconciliation coming from New York, Paris or Brussels, it may be pointless and 'out of touch' to plan to 'begin with' reconciliation. In any case, reconciliation seems to be a spontaneous need for the population. It is, however, fundamental to highlight that reconciliation does not naturally become easier with time. There is some truth in the statement 'justice delayed is justice denied' – the pain, frustrations and anger that are the legacy of violence will only grow and not diminish if they are not addressed *at some point* (the betrayal felt by survivors of Apartheid-era violations who are 'still waiting', as Mueller-Hirth demonstrates in Chapter 7, are a case in point). We must therefore consider the feasibility of implementing reconciliation gradually, in parallel with other reconstruction activities – neither before, nor simply afterwards. This delicate balance leaves one question unanswered: when does a protagonist emerge ready to make efforts towards reconciliation?

Ripeness: not too soon

William Zartman (2000) has emphasized the significance of ripeness in the area of negotiation (see also Rios Oyola's Chapter 4 in this volume). This issue should also be addressed with respect to reconciliation. As in any negotiation, parties become involved in a reconciliation process when they have to do so. They only join together in this way if each party's efforts to achieve a unilaterally satisfactory result are blocked and if the parties feel trapped in a costly predicament. Similarly, the notion of 'mutually hurting stalemate' is relevant in cases of a rapprochement between former enemies. Parties that find themselves locked in a situation that is painful for both (although not necessarily to the same extent or for the same reasons) seek a way out that can indeed become the starting point of a reconciliation process.

The post WWII Franco-German case is telling in this regard. In a devastated Europe, the decision to work towards a rapprochement was not a matter of altruism, but was rather in both French and German national interests. The complete and radical character of Germany's defeat explains its crucial need for political rehabilitation and a restoration of sovereignty. Moreover, to German leaders, the economic future of their country was an additional reason to favour as quickly as possible the normalisation of relationships with their neighbours. In this particular context, rapprochement with France was depicted as a real necessity. As for France, it was also a question of needs must. After the war, the status of France had been called into question. Its economy was reduced by half, its infrastructures were devastated, its demography was compromised by the human cost of the conflict, and it was on the point of losing its colonies. Both countries needed one another. Aside from these domestic issues, the configuration of the broader international system was decisive in stimulating a rapprochement. Among all the political, economic and security considerations that promoted the transformation of relations between 'hereditary enemies', one was particularly significant: the existence of a common enemy in the USSR, and therefore the external, mostly American, support for rapprochement. In such circumstances, French and German leaders perceived an undeniable 'mutually hurting stalemate' and considered the European Coal and Steel Community (ECSC), a six-nation western European economic union, as their way out.

The Rwandan case is also particularly telling (also see Chapter 8 on this volume, where Richard Benda explores transitional temporalities and the intergenerational transmission of guilt in post-genocide Rwanda). When the Rwandan Patriotic Front came to power in 1994, the objective was not reconciliation but justice. The message repeated in the country and abroad was that Rwanda had to erase the culture of impunity for identity-based collective violence. Eight years later, the Rwandan president Paul Kagame started focusing on reconciliation. Three main reasons justified this shift. Firstly, justice was in fact impossible, since the judiciary system was totally ruined after the genocide and the number of people who had to be tried and sentenced was simply unmanageable. Secondly, justice could result in the revelation of embarrassing truths.

Justice meant dealing with crimes committed on each side, including those which were committed in the Congo after July 1994. Thirdly, most international donors strongly emphasized the need for reconciliation in Rwanda. This was particularly true of Belgium.

These examples show that, at the political level, ripeness directly depends on the national and international context. There are many other cases which signal that in the absence of a mutually hurting stalemate, the pain is not strong enough to drive the parties towards reconciliation. Situations which have remained unchanged for years, such as that in Cyprus, are emblematic in this regard. In cases where spoilers considerably benefit from intractable conflicts, ripeness in the eyes of the population cannot lead to effective change on the ground. However, the absence of ripeness is not a valid reason for inaction. In cases such as Israel/Palestine or the Democratic Republic of the Congo (DRC), certain groups are able to conceive of a rapprochement with *the other*. Admittedly, without political support 'from above', their efforts may be insufficient to influence the whole population. Nevertheless, the existence of '*pause conflict zones*' (in contexts that cannot yet be qualified as 'post-conflict zones') can constitute decisive niches once there is a real eagerness for reconciliation. They create the social infrastructure which political leaders can count upon if they decide to take the risk – or the chance – of a rapprochement.

Tempo: what is the appropriate pace?

All these elements show how delicate the issue of time management is in bringing about reconciliation. While institutional and legal measures can be taken relatively quickly after the end of the conflict, so-called reconciliation policies must not come too soon. If so, they are ineffective and even sometimes detrimental to peace. The intensity of emotions that characterize post-war settings indicates that the end of the hostilities is only the starting point of a long-term challenge. The significance of passions (Hassner 2005) such as humiliation, anger, hatred or sorrow explains to a considerable extent why 'negotiating reconciliation' cannot be reduced to any other form of bargaining. Far beyond the balance of power (which remains critical), the understanding of post-war negotiations requires insight into psychological processes traditionally overlooked in international relations. How can the resistance of affected groups be understood without taking into account the wounds of those who were directly or indirectly (if considering subsequent generations) affected by the war? How can cases like Iraq, Afghanistan, Syria or Libya be comprehended if there is not a realization that these harms have not only left open but also festering wounds?

An invasive past

It might be counter-productive to formulate general considerations without taking into account the potential perils of context-insensitive attempts at reconciliation. Having said that, there are clearly experiences common to those who have been affected by violent conflict across different contexts. From that viewpoint, the following schema summarizes the potential evolution that may or may not occur in the aftermath of wars and/or mass atrocities. In any post-conflict situation, the sequence 'past – present – future' says a lot about people's concerns and expectations. Far from being watertight, these three notions fluctuate and often overlap. However, the way people describe their past, their present and their future reveals whether a rapprochement might or might not occur. At the very end of the hostilities, the importance of the past is uppermost in people's minds: the present is relatively less significant, and the future even less so. The process of reconciliation is intended to reverse this trend, leading to a situation where energy is forward and not backward-looking (Zartman 2005). According to Hayner, reconciliation can be understood as a way to build 'relationships today that are not haunted by the conflicts and hatreds of yesterday' (2000: 161).

From that perspective, the critical question is not only 'What happened?', but also – and above all – 'What shall we do with the past?'. Similarly, one of the most fundamental issues is not whether to remember *or* forget, but how to remember *and* forget in order to move forward. Events such as particularly traumatic violations of human rights can remain unexpressed for a period of time – a period that psychoanalysts often call 'latent' (Weinrich 1999: 189). Some specialists refer to a period of twenty-five years (Rimé 2013). However, there does not appear to be any standard time period in this matter.

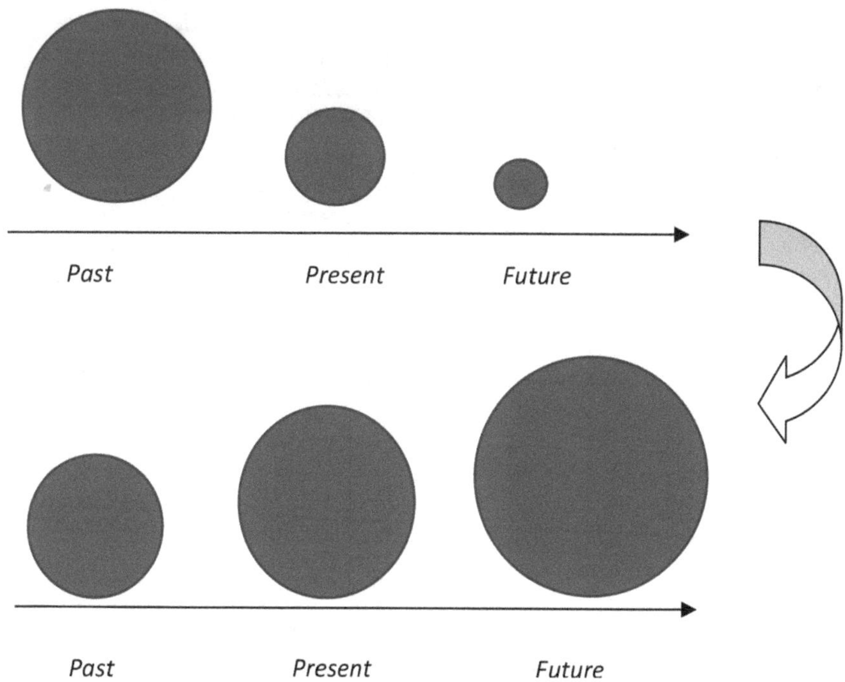

Figure 2.1 Time in Reconciliation

After World War II, the variety of reactions by survivors of the Nazi atrocities is striking. Some of them directly chose to speak up even though they were rarely believed or understood. Others remained silent until their death. Between these two extreme cases (immediately *versus* never), some survivors started elaborating about the past after years, or even decades. Psychologically speaking, looking back was dangerous, if not fatal (Rosenblum 2009). According to clinicians, survivors who underwent torture are confronted with a specific difficulty, which is the inexpressibility and 'unsharability' of physical pain since 'physical pain does not simply resist language, but actually destroys it' (Scarry 1987: 4).

Beside these psychological and physical dimensions, political and social considerations explain why some memories can remain unexpressed for a rather long time. Political transitions, dictatorial regimes or new waves of violence often 'freeze' unauthorized expressions of the past. However, these obstacles to transmit explicit narratives of the past do not prevent the transmission of emotions such as anger, hatred, resentment, grief, shame or guilt. This particular transmission directly concerns family circles and is intergenerational (Thompson 2002).

Thinking in terms of intergenerational processes is not new. In Greek tragedies, already, old debts were only paid after three generations (Salles 2012). However, the notion of a transgenerational transmission of trauma (Alexander et al. 2004; Volkan 2006) is debatable. Some sceptical voices argue that children of survivors do not systematically inherit their injuries (Winter 2017). However, even if we do not consider that 'traumatic memories' are literally passed on to one's children and other loved ones, it is difficult to deny that the younger people who live in the presence of survivors of war and genocide feel the emotional pull of their elders' suffering to such an extent that they can appropriate the memories of the first generation as their own (Argenti and Schramm 2009; also see Benda, Chapter 8).

Numerous studies in social psychology demonstrate that the second and third generations remain preoccupied with continuing feelings of guilt or victimization, especially when past injustices have not been adequately addressed by their parents' generation (Klar and Branscombe 2016). The Armenian–Turkish case is emblematic in this regard. The persistence of intense emotions has had specific consequences for the negotiation process. It limits the potential concessions made on each side and partly explains why certain issues have remained non-negotiable for decades. Resistance to recognizing an atrocity can be so strong as to be qualified as 'irrational'. One way of understanding the dynamics that led to this kind of impasse is to consider the extent to which the process may be 'haunted' by the past. In the aftermath of mass atrocities, it is often meaningful to refer to the notion of 'ghosts'. As a former American ambassador to Iran once explained, parties should always realize that after a war, there are always many more protagonists at the table than one might initially have thought (USIP 2009).

These words are echoed tragically by the experience of Révérien Rurangwa. After witnessing the massacre of forty-three relatives, this Rwandan survivor was disfigured and left for dead. When he decided to take up his pen to express the 'infinite sadness of the irreparable' (Rurangwa 2007: 146), he described himself as a 'prisoner of his ghosts and his anxieties', explaining: 'my body, my face and the sharpest aspect of my memory bear the scars of the past – and will continue to do so until the end of my life' (2007: 14). The same experience is depicted by several survivors. When the Auschwitz survivor Charlotte Delbo wrote her memoir of the war, she explained that the camp existed for her in a 'perpetual present'. To her, Auschwitz was 'there, fixed and unchangeable' (1970: 13).

Cadence: not too quick

The emotional weight of the past can take many forms, from silence to resentment. Can it gradually be lifted? The transformation of the representations that parties have of the past is an ongoing process that generally lasts for decades. As the former Polish President Bronislaw Komorowzki said after talks with the former Russian President Dimitri Medvedev, 'the path towards reconciliation is a marathon, not a sprint' (quoted by Dempsey 2010).

Beside the psychological dimension of the issue, it is worth underlining the influence of leadership on the pace of the rapprochement. The attitude of leaders is critical both to launching *and* maintaining the process. Case studies show that things will go more smoothly if the rapprochement is advocated by a person who has accomplished heroic actions against the enemy with whom reconciliation is being sought. This person then asks the population to undergo a transformation that he has undergone himself – i.e., overcoming resentment towards the former enemy. For instance, the historical legitimacy of Charles de Gaulle probably helped the French to change their views about the Franco-German relationships. A similar point can be made with respect to Nelson Mandela in South Africa.

In the Franco-German case, it took two generations to change the interpretation given to the Battle of Verdun (where nearly 800,000 French and German soldiers were killed or wounded in an inconclusive fight over a few square miles of territory between February and December 1916). Initially depicted as a symbol of the hatred and ferocity of the adversaries, Verdun gradually came to be described as a 'common tragedy' and a 'shared martyrdom' on both sides of the Rhine. It is, moreover, worth drawing attention to the gap – in terms of tempo – between official and social processes. Officially, the French and German authorities highlighted the need to face up to the past just over ten years after the end of the Second World War. Charles de Gaulle and Konrad Adenauer wrote speeches, held commemorations, and made symbolic gestures to show consideration of each other's past. By contrast, a systematic analysis of the education-related activities organized by the Franco-German Youth Office from 1963 to 2001 shows that painful questions relating to the past were not openly addressed before the early 1990s (Delori 2016).

In case of South Africa, the gap between institutional time and individual time is similar. One of the main shortcomings of the Truth and Reconciliation

Commission was precisely that reconciliation in such a context was an ambitious but dubious goal. Individual reconciliation has a different time scale from that of any commission (Hamber and Wilson 2002). Individual healing is a process that advances at its own pace, which cannot be pushed or programmed. In the space of a few years, truth commissions could only create the conditions in which reconciliation might eventually occur (Garton Ash 2003: 416). Van Roekel's work (Chapter 5, this volume) highlights this gap between lived experiences of time and 'transitional justice time' (Hinton, this volume) by exploring senses of time and conceptions of justice of victims and indicted military officers in Argentina. The disjuncture between institutional and individual/communal levels is also illustrated by Igreja's study of the multiple temporalities of accountability in Mozambique (Chapter 6, this volume): serious violations had never been subjected to official and time-limited processes of accountability (such as they are encapsulated in transitional justice mechanisms), but embodied accountability emerged from the disruptive intrusions of spirits that operate irrespective of the passage of time.

Closure: when is it over?

The open-ended nature of the process is confirmed by all case studies. In 2012, by way of an illustration, the Estonian president, Toomas Hendrik Ilves, explained that, more than sixty years later, Estonia was 'still waiting for apologies' from the Russian authorities (*Le Monde*, 12 April 2012). Speaking about the 'illegal annexation' of his country by the Soviet Union after WWII until 1991, and more specifically about the deportation of 20,000 Estonians to Siberia by the NKVD (Soviet political police) in 1949, he expressed regret that nobody had 'ever apologized' for these crimes. In another context, negotiations between China and Japan are constantly at risk from the explosive weight of the past. To give a final example, the ongoing negotiations between Japan and North Korea have always failed to achieve closure regarding the Japanese who were kidnapped in the 1970s and 1980s, and who were used as spies for Pyongyang (*Le Monde*, 31 May 2014). Each of these examples – and there are many others – reminds us that the transformation of relationships between former enemies is all about time.

An irrevocable past

The duration of the process is confirmed by the experience of a Rwandan survivor, actively involved in a local reconciliation process: 'I took the time to hate everybody. It took me ten years. I needed this time for hatred. Now I can think about reconciliation.'[3] In listening to this sort of comment, many observers emphasize the impact of a supposed 'natural time for healing'. Thus, in April 2014, the Canadian government considered that a period of two decades was symbolically enough to 'recover' and turn the page of the Rwandan genocide (Montgomery 2014). Likewise, religious leaders and NGO workers often quote a famous passage of the Bible to corroborate that time heals the wounds:

> There is a time for everything, and a season for every activity under the heavens: a time to be born and a time to die, [...] a time to kill and a time to heal, a time to tear down and a time to build, a time to weep and a time to laugh, a time to mourn and a time to dance, a time to scatter stones and a time to gather them, [...] a time to be silent and a time to speak, a time to love and a time to hate, a time for war and a time for peace.
>
> (Ecclesiastes 3: 3–8)

This quotation is usually understood as confirming the healing impact of the mere passage of time. However, this emphasis on a quasi-magical effect of time has to be questioned in at least two ways. First, who can determine whether ten, twenty, fifty or a hundred years is the right time period? How can we fix in advance a standard pace of healing? Secondly, how can one argue in favour of such natural healing processes when in most cases there is evidence that wounds remain? As numerous testimonies remind us, the idea of a natural time for healing does not systematically correspond to reality – far from it. Two examples are particularly emblematic of this oversimplification of survivors' experiences.

The first is that of the philosopher, Jean Améry, who was a Resistance fighter during the Second World War. Born in 1912 in Austria, Jean Améry fled to Belgium and joined the Resistance. Caught and tortured by the Gestapo, he was sent to several concentration and extermination camps including Auschwitz. Upon liberation from Bergen-Belsen, Améry started writing about his experience. Far from his burden being lifted over time, Améry committed suicide in Salzburg in 1978. In his book *Beyond Guilt and Atonement*, his words criticize the official narrative of reconciliation and emphasize the permanence of his wounds: 'What had happened had really happened' and 'the fact that this happened cannot be taken thoughtlessly'; 'nothing has scarred over, and the wound that was almost healed always reopens and suppurates' (2005: 17–20). The way in which Jean Améry insists upon his right to resentment shows the need to go beyond the traditional vision of resentment as a backward-looking emotion or as a sign of a moral failure (Brudholm 2008). He refuses to hail the Franco-German rapprochement as a success, since his moral understanding of what happened is that it was absolutely unacceptable; something with which there can be no reconciliation (Brudholm and Rosoux 2013). Améry's is not an isolated voice reminding us of how long resentment can last, and how intense it can be, in the aftermath of mass atrocities.

In the Rwandan context, Rurangwa's words are hard-hitting: 'Survivors are troublemakers. We bother everybody with our pain, our tragedies, our dead ones and our rotten vivid memory. (...) We feel that it is better not to tell (...). The only narrative we are allowed can be summed up in two words: 'Reconstruction, Reconciliation'. These watchwords – as nice as they are *per se* – smother the survivors' distress cries' (131). Like Améry, Rurangwa attempts to share a disturbing reality. 'Too painful to tell, too painful to listen to. Too long. Too everything' (134).

These accounts raise fundamental questions regarding closure in post-conflict situations. One of them was once summarized by a German student of mine, who came to me after a seminar and asked, movingly: 'How long?'[4] Her two-word question meant 'How long do we Germans have to apologize?'; 'How long do we have to deal with a guilt that is not ours?'; 'How long will we have to justify ourselves, compensate for and/or recall the events of the past?'. The answer given by the German Minister of Foreign Affairs, Frank Walter Steinmeier, confirms the open-ended character of the process. As he admitted: 'It is very dangerous for a politician to say that we have succeeded. We have to bring evidence of this reconciliation [between Germany and its former enemies] all the time' (*La libre Belgique*, 18 February 2014).

These questions also point to the unpredictable nature of subjective timing and reactions to the reconciliation process. The need to acknowledge again and again the crimes committed in the past stresses the discrepancy between various temporalities. Two illustrations at least should be considered. The first is the patience shown by certain individuals who have been particularly determined to get some form of justice. The Inuit leader and survivor of Canada's residential schools programme, Peter Irniq, is a striking example of this unusual persistence. As he explains, 'I come from a family of seal hunters. We are used to sitting patiently on the ice for hours, at temperatures way below freezing, waiting for a seal. [...] *No matter how long it takes*, we will wait for justice' (quoted in the ICTJ Annual Report 2005: 11). A second instance shows not the long-term determination of those who call for justice, but rather the genuine diversity in the length of time taken for attitudes to change. In this regard, the South African TRC provoked a surprising variety of reactions. Among them, the attitude of Siphiwo Mtimkulu's mother is revealing in terms of timing. Mtimkulu was a 22-year-old student leader who was shot and burned in 1982 by the security police under the apartheid government. Fifteen years later, Gideon Nieuwoudt, who was a colonel in the security police and responsible for the torture, poisoning and death of numerous black activists, including Steve Biko, met with the Mtimkulu family seeking their forgiveness. The response given by Siphiwo's mother was uncompromising: 'It is too late to us. To God, it is never too late. But to me, it is too late. You came too late.'[5]

As we look at the many examples, we realize that the way in which past experiences are re-lived ultimately depends on each individual's attitudes, emotions and energy, and crucially on issues around redistribution, compensation and fairness (see for example Mueller-Hirth's Chapter 7 on South African victims' experiences of waiting for reparations, this volume). International and intercommunity conflicts provoke an infinite series of individual fires that need to be extinguished one by one. The response to past atrocities is ultimately an individual one.

Irreversible: outside of time

The two previous sections focused on the contextual and individual variables. The issue of closure can hardly be understood without paying attention to the nature of the past violence. To show the significance of this third variable, it is useful to compare the failed Treaty of Friendship between France and Algeria with the successful Elysée Treaty between France and Germany. Why was closure impossible in one case and not in the other? Why can Franco-Algerian relationships apparently not be 'normalized'?

Since the end of the Algerian war in 1962, French and Algerian authorities have frequently referred to the Franco-German case as a model for moving forward. In 1983, Chadli Bendjedid undertook the first ever state visit by an Algerian president to France and directly evoked the Franco-German relations: 'Why couldn't there be identical relations between France and Algeria?' (*Le Monde*, 6–7 November 1983). In 2001, the former French President Jacques Chirac stressed the same precedent:

> The weight of the past finally fades with time. The weight of the past was much more difficult to erase between Germany and France (...). The dispute was age-old, considerable and added up to millions and millions of dead, during successive wars. Thus, I am deeply convinced that the relation between France and Algeria is in the nature of things (...) and that it can develop.
>
> (Algiers, 1 December 2001)

From that perspective, it was not surprising that Jacques Chirac explicitly called for an 'Elysée Treaty in the Franco-Algerian style' (*Le Point*, 19 August 2004).

However, the failure of the negotiations leading to a Friendship Treaty shows that the Franco-German model could not be easily replicated. The intensity of the domestic resistances in both countries reveal that the wounds described by various groups on both sides of the Mediterranean Sea (*pieds-noirs*, former moudjahids, former French combatants, *harkis* – i.e. Muslims who fought alongside the French against their fellow Algerians) remain open. The case of the *harkis* is particularly telling. Following the French withdrawal, up to 150,000 *harkis* were slaughtered in Algeria.[6] More than 40,000 *harkis* were able to escape to France after the war, but they were badly treated once they arrived. Most of them described a double betrayal (not only by Algeria but also by France), and considered themselves as second-class French citizens. The descendants of the *harkis* nowadays insist on the long-term impacts of this double rejection: financial distress, a high unemployment rate and the high frequency of suicides in their families. To them, this issue is far from being closed. During the negotiation process, Jacques Chirac suggested that the *harkis* be mentioned at the moment of the signature of the treaty, while Algiers did not want to hear anything about these traitors.

The *harkis* are not the only group still struggling with the meaning of the past. The *pieds-noirs* (or *black feet*, as the white French former colonists in Algeria were called) also depict themselves as 'historical victims of social exclusion'. Testimonies are abundant in this regard: 'we are actually the losers, we have been manhandled, misled, humiliated, tortured, imprisoned, broken, rejected, caricaturized'; 'we are a dead people. Without geography, there is nothing left' (Baussant 2002: 424 and 433). The feeling is the same in the mind of former French combatants who do not feel like former combatants of WWII because people are not interested in them. A lot of them lament that their fight did not make any sense, 25% of them considering that their stay in Algeria was actually useless (Jauffret 2000: 329). As all these reactions show, one of the features of the Franco-Algerian case is the lack of psychological closure.

The research carried out so far indicates that this situation largely results from the nature of the violence that was committed in the past. On both sides of the Mediterranean, thousands of families are still trying to cope with the disappeared. As many other case studies show (from Argentina to Sri Lanka and South Africa), the lack of bodies to be buried creates another temporal continuum. Families of the disappeared can hardly start a mourning process without any trace of their loved ones. The uncertainty about their fate and the absence of burial sites prevent them to move forward. Some explain that time has been stopped. Others attempted to move on but find themselves systematically blocked and sent back to the period of the disappearance – again and again, year after year. So many interviewed people keep repeating: 'it is as if it occurred yesterday', confirming that the violence of the past can literally 'contaminate any possibility of a new present' (Mouchard 2007: 108).

Beyond the specific issue of the disappeared (that is directly related to the use of torture during the Algerian war), it is worth considering the process of othering that led to past violence. In the framework of the Franco-German wars, the other was the *enemy* to fight. In the colonial context, the other – as depicted by the colonial authorities – was a backward *child* to be educated and/or a *barbarian* to be exploited. These representations do not have the same long-term effects on the affected population.

For centuries, the German enemy was a basic component of French national identity and vice versa. Traditional songs on both sides of the Rhine revealed not only a constant competition between national powers, but also a genuine hatred between the nations. However, this hatred is not sufficient to depict the true depth of Franco-German relationships. One of the key facets of this case lies in the ambiguous fascination that captivated both peoples. Interestingly enough, this fascination was not incompatible with the detestation of the enemy. In fact, respect for French culture was commonplace among the German elite. Likewise, a long tradition of French intellectuals and artists expressed their admiration for German writers and composers. This uneasy symmetry, made up of a mixture of hatred and admiration, would be decisive in creating the favourable conditions for a post-war rapprochement.

The Franco-Algerian case is totally different. First, colonialization can hardly be characterized as a period of reciprocal admiration. Scorn and humiliation were felt on a day-to-day basis. Secondly, the nature of the war was very different. Far from being a war between similar combatants on both sides (as in the case of Verdun during WWI for instance), the fighting between the French army and the *fellagha* cannot be qualified as symmetrical. Thirdly, the war in Algeria ended after a negotiated agreement (the Evian Accords in 1962), and not after a crushing defeat by one of the parties. From that perspective, the notion of winners/losers is obviously less relevant than in other circumstances. Therefore, it seems appropriate to question the notion of friendship in such circumstances. Is friendship possible – and even, necessary – in all settings?

Conclusion

The objective of this chapter has been to show that beside the plurality of interests that characterize any post-conflict situation, we must also consider a multiplicity of time frames. Survivors, criminals, families, bystanders, outsiders, returnees, former combatants, religious associations, international donors all operate within their own tempo and, within each of these groups, everyone may react in a unique and largely unpredictable way. Part II of this book in particular explores these multiple temporalities in a range of case study countries. This multiplicity of temporalities forces us to question enthusiastic, and, often, premature calls for reconciliation. To most official representatives meeting in Brussels and Washington, it is now 'time to move on' in Rwanda or Bosnia. As former US Secretary of State, Hilary Clinton, clearly explained in Kinshasa, '[w]e can go work with people who are willing to forget the past and focus on the future. We are not going to work with people who are looking backwards, because that's not going to get us where we want to go' (11 August 2009). This eagerness to move forward is understandable. However, it does not easily resonate with those victims who do not feel ready to take up the challenge, at least not in the short term. This gap is highlighted by a Rwandan survivor left alone in her village, who slowly replied to some ambitious outsiders: 'I can live with them. Don't ask me more. Don't ask me too much'.[7]

These few words remind us that we must be pragmatic, and must consider the real situation in the areas affected: who are the parties? What are their emotions, energy, hopes, and to what are they most resistant? What would be a realistic target, given that the level of hatred is high and the level of trust quasi non-existent? Rather than being prescriptive, it might be useful to question our own assumptions. Is reconciliation a search for consensus about the past, or is it a way to bring together different viewpoints of it? Is the aim to rapidly develop relationships of trust, or is it to deal resolutely with legitimate resentment? Is it to distinguish between 'good' (resilient and forgiving) victims and 'bad' victims (expressing signs of hatred) or to define a new social contract?

These questions highlight the need for arduous and possibly never-ending negotiations at all levels, from official negotiations at the macro level to daily inter-individual negotiations at the micro level. From that perspective, I will always remember the eyes of a Colombian woman who told me, tragically: 'Don't touch my hatred. That is the only thing that's left. They took all I had – except for my hatred.'[8]

Rather than expecting a process that entails forgiveness and harmony, I would insist on the importance of setting achievable aims and being realistic in terms of timing. In post-conflict societies, decision-makers cannot avoid dealing with contradictory interests, explosive emotions and multiple time frames. It is if – and only if – the real duration of these multi-layered negotiation processes is taken seriously that one can finally see an end and a beginning.

Notes

1 'The dead are invisible, but they are not absent'. Victor Hugo, 19 January 1865, *Actes et paroles. Pendant l'exil* (1865), in 1985. *Œuvres complètes.* Paris: Robert Laffont, 65.
2 Two main kinds of data were combined to dissect the relevant processes in each scenario. First, a systematic corpus of official speeches allows for a description of the evolution of the leaders since the end of the hostilities. Second, a comprehensive gathering of narratives depicts the reactions of individuals directly affected by a violent past. This chapter is based on a study that was partly published in Rosoux and Anstey (eds) 2017. *Negotiating Reconciliation in Peacemaking*. Rotterdam: Springer. I wish to thank the Max Planck Institute Luxembourg for supporting this project. I also thank the two encouraging and meticulous editors of this book.
3 Louvain-la-Neuve, 18 October 2005.
4 Louvain-la-Neuve, 11 February 2014.
5 See the documentary *Where the Truth Lies*, by Mark J. Kaplan, Bullfrogfilm, 1999.
6 The figures still vary according to the sources.
7 Kigali, 7 April 2010.
8 Paris, 23 April 2013.

References

Alexander, Jeffrey *et al.* 2004. *Cultural Trauma and Collective Identity.* Berkeley: University of California Press.

Améry, Jean. 1995. *Par-delà le crime et le châtiment. Essai pour surmonter le mal.* Paris: Actes Sud.

Argenti, Nicolas and Schramm, Katharina (eds). 2009. *Remembering Violence: Anthropological Perspectives on Intergenerational Transmission.* New York: Berghahn Books, 1–40.

Barsalou, Judy. 2005. Trauma and Transitional Justice in Divided Societies. *Special Report of the United States Institute of Peace (USIP)*, 135.

Baussant, M. 2002. *Pieds-noirs. Mémoires d'exil.* Paris: Stock.

Brudholm, Thomas. 2008. *Resentment's Virtue: Jean Améry and the Refusal to Forgive.* Philadelphia: Temple University Press.

Brudholm, Thomas and Rosoux, Valerie. 2013. The Unforgiving: Reflections on the Resistance to Forgiveness after Atrocity. In Alexander Hirsch (ed.), *Theorizing*

Post-Conflict Reconciliation: Agonism, Restitution and Repair. London: Routledge, 115–130.

Delbo, Charlotte. 1970. *Auschwitz et après. I*. Paris: Editions de Minuit.

Delori, Mathias. 2016. *La réconciliation franco-allemande par la jeunesse. La généalogie, l'événement, l'histoire (1871–2015)*. Brussels: Peter Lang.

Dempsey, John. 2010. 'Old foes see reasons to get along', *International New York Times*, 8 December.

Garton Ash, T. 2003. The Waters of Mesomnesia. In Anne-Marie Le Gloannec and Aleksander Smolar (eds), *Entre Kant et Kosovo. Etudes offertes à Pierre Hassner*. Paris: Presses de Sciences Po, 405–419.

Hamber, Brandon and Wilson, Richard. 2002. Symbolic closure through memory, reparation and revenge in post-conflict societies. *Journal of Human Rights* 1(1), 35–53.

Hassner, Pierre. 2005. La revanche des passions. *Commentaire*, no.110, 299–312.

Hayner, Patricia. 2000. *Unspeakable Truths: Confronting State Terror and Atrocity*. London: Routledge.

Human Rights Watch (HRW). 2012. Q&A about Jospeh Kony and Lord's Resistance Army. Retrieved 12 October 2017 www.hrw.org/news/2012/03/21/qa-joseph- kony-and-lords-resistance-army.

International Center for Transitional Justice (ICTJ). 2005. A Voice for Victims. Annual report 2004/2005. Retrieved 12 October 2017 www.ictj.org/sites/default/files/ICTJ_AnnualReport_2004-5.pdf.

Jauffret, J.-C. 2000. Soldats en Algérie, 1954-1962. *Expériences contrastées des hommes du contingent*. Paris: Autrement.

Klar, Yechiel and Branscombe, Nyla R. 2016. Intergroup reconciliation: Emotions are not enough. *Psychological Inquiry* 27(2), 106–112.

Kwinter, Sanford. 2002. *Architectures of Time: Toward a theory of the Event in Modernist Culture*. Boston, MA: MIT Press.

Langer, Lawrence L. 1991. *Holocaust Testimonies: The Ruins of Memory*. New Haven, CT: Yale University Press.

Montgomery, Sue. 2014. 'Rwanda: 20 years later: The burden of survival'. *The Gazette*, 1 April.

Mouchard, Claude. 2007. *Qui si je criais…? Œuvres-témoignages dans les tourments du XXè siècle*. Paris: Editions Laurence Teper.

Paine, Thomas. 2008. *Common Sense: Address to the Inhabitants of America*. Gutenberg ebook 147.

Rimé, Bernard. 2013. Mémoire collective et commémorations: processus psychosociaux et fonctions psychosociales. Conference paper given at UC Louvain, 6 June.

Rosenblum, Rachel. 2009. Postponing trauma: The dangers of telling. *The International Journal of Psychoanalysis* 90(6), 1319–1340.

Rurangwa, Révérien. 2007. *Génocidé*. Paris: J'ai lu.

Salles, Alain. 2012. 'Ubris, catharsis et "troika"', *Le Monde*, 11 May.

Scarry, Elaine. 1987. *The Body in Pain: The Making and Unmaking of the World*. Oxford: Oxford University Press.

Schaap, Andrew. 2007. The Time of Reconciliation and the Space of Politics. In Scott Veitch (ed.), *Law and the Politics of Reconciliation*. Burlington, VT: Ashgate, 9–29.

Sissons, Miranda and Al-Saiedi, Abdulrazzaq. 2013. A Bitter Legacy: Lessons of De-Baathification in Iraq. International Center for Transitional Justice. Retrieved 12 October 2017 http://ictj.org/sites/default/files/ICTJ-Report-Iraq-De-Baathification-2013-ENG.pdf.

Thompson, Janna. 2002. *Taking Responsibility for the Past: Reparation and Historical Justice.* New York: Polity.

UN Security Council. 2004. 'The rule of law and transitional justice in conflict and post-conflict societies'. Report of the Secretary-General to the Security Council, 24 August (S/2004/616).

UN Security Council. 2011. 'The rule of law and transitional justice in conflict and post-conflict societies'. Report of the Secretary-General to the Security Council, 12 October (S/2011/634).

United States Institute of Peace (USIP). 2009. *Guiding Principles for Stabilization and Reconstruction.* Washington, DC: USIP Press.

Volkan, Vamir. 2006. 'The next chapter: Consequences of Societal Traumas'. Keynote address at Memory, Narrative, and Forgiveness, University of Cape Town, 22–25 November. Retrieved 10 May 2017 www.vamikvolkan.com/The-Next-Chapter%3A-Consequences-of-Societal-Trauma.php.

Weinrich, H. 1999. *Léthé: Art et critique de l'oubli.* Paris: Fayard.

Winter, Jay. 2017. Commemorating Catastrophe: The First World War, Europe's Suicide? In Etienne François and Thomas Serrier (eds), *Europa. Lieux de mémoire européens.* Paris: Les Arènes, 35–53.

Zartman, William I. 2000. Ripeness: The Hurting Stalemate and Beyond. In Paul Stern and Daniel Druckman (eds), *International Conflict Resolution after the Cold War.* Washington, DC: National Academy Press, 225–250.

Zartman, William I. and Kremenyuk, Victor (eds), 2005. *Peace versus Justice Negotiating Forward- and Backward-Looking Outcomes.* Oxford: Rowman & Littlefield.

3 Transitional justice time

Uncle San, Aunty Yan, and outreach at the Khmer Rouge Tribunal[1]

Alexander Laban Hinton

Introduction

Uncle San lives suspended in time. He is caught in a state of traumatic dysfunction, unable to escape from a violent past that afflicts him with nightmares. Uncle San stands as the image of the Cambodian everyman or of Cambodia itself. And he lives in transitional justice time. It is a timescape of the 'other', at once helped and remade.

Who is Uncle San? He is the protagonist in a 34-page booklet, entitled *Uncle San, Aunty Yan, and the KRT* [Khmer Rouge Tribunal], the pages of which are divided (on the left page) by visual representations accompanied (on the right-hand page) by explanatory text, which was produced in 2008 by the Khmer Institute of Democracy (KID), a non-government organization in Cambodia conducting outreach for the Khmer Rouge Tribunal (KRT), officially known as 'The Extraordinary Chambers in the Courts of Cambodia' (ECCC).

After several years of negotiation, this UN-back hybrid court was established in 2003 to try the surviving leaders of the Khmer Rouge. Upon taking power in Cambodia on 17 April 1975, this group of Maoist-inspired revolutionaries set out to launch at 'Super Great Leap Forward' that would outdo all other communist revolutions, even that of their close socialist ally, China. By the time of its downfall on 6 January 1979, the policies the Khmer Rouge implemented, ranging from collectivization and forced labour to increasingly frequent purges of suspected enemies, had resulted in the death of perhaps 1.7 to 2.2 million of Cambodia's 8 million inhabitants, almost a quarter of the population, due to starvation, overwork, malnutrition, or execution.[2]

The ECCC, whose mandate is to try the 'senior leaders' and those 'most responsible' for the atrocities that took place during this period, began operation in 2006. The court includes international and Cambodian officials in all major offices, including co-lawyers, co-prosecutors, and co-judges.[3] It is one of a growing number of 'hybrid' tribunals established in the 2000s (other hybrid tribunals include Kosovo, Sierra Leone, Timor Leste, and Lebanon) to offset some of the problems that emerged with the *ad hoc* tribunals of the 1990s (the International Criminal Tribunal for the former Yugoslavia and

Rwanda), particularly in terms of cost, duration, and proximity to the populations involved in given conflicts.

Hybrid tribunals, like their *ad hoc* counterparts and other modes of redress such as truth and reconciliations commissions, are often referred to as transitional justice mechanisms. The field of transitional justice emerged in the late 1980s and early 1990s as the Cold War was ending and a number of places, ranging from Southern Cone states to Eastern European countries, were grappling with recent violent pasts while seeking a way forward, what was often referred to as 'democratic transition'.

The term 'transitional justice' emerged in the post-Cold War 'new world order', as criminal tribunals, truth and reconciliation commissions, memorialization and reparation efforts, and institutional reforms, emerged as favoured mechanisms for providing some sort of redress and sense of justice that would enable these countries under transition to move from troubled pasts to better futures.[4] Such transitional justice mechanisms directly grapple with the legacies of the past and are often said to have a preventative dimension (see the editors' introduction to this volume), diminishing the potential for the recurrence of violence by educating the population and seeking to 'combat impunity', promote the 'rule of law', reveal 'the truth', and educate the larger populace.

More broadly, this chapter argues that transitional justice mechanisms are based on a particular concept of time, one that is manifest in the booklet *Uncle San, Aunty Yan, and the Khmer Rouge Tribunal* and is part of a larger 'transitional justice imaginary',[5] or set of interrelated discourses, practices, and institutional forms that, through performance, help generate a sense of shared belonging among a group of people – in this case the transitional justice community (which is itself is part of the larger 'international community') in the broadest sense.[6] Such imaginaries may be gleaned from a variety of sources, ranging from ritual practices to myths and, as in this case, stories.

The transitional justice imaginary is not monolithic, and varies across localities and individuals. Nevertheless, I would argue that it can be found in most transitional justice contexts, including the Khmer Rouge Tribunal, as the members of the 'international community' and local elites constitute themselves through the assertion of a transitional justice imaginary – one that may differ significantly from local vernaculars (an examination of this issue is beyond the scope of this essay but is addressed in the book on the tribunal that I am writing).

The transitional justice imaginary is normative (i.e., it is associated with certain truth claims and moral-laden assumptions), performative (i.e., through its enactment, people constitute an imagined community), and productive (i.e., the imaginary produces certain subject positions and types of being). The imaginary is also characterized by a particular temporality, what I am calling 'transitional justice time', premised on a value-laden pre-post-conflict state of conflict and teleological movement between them).

More specifically, implicit within transitional justice time is a highly normative concept of past and present. Violent pasts are delimited and narrowed, erasing historical complexities and suggesting an essentialized notion of regressive being, epitomized by phrases such as 'failed states' or indexical registers equating a country with violence and death (for example, the frequently juxtaposing of countries like Cambodia next to images of skulls). This foreshortened vision of the authoritarian past is set against an imagined liberal democratic future, with transitional justice as the mechanism of teleological change. This splitting of past and future, mediated by a liminal present, is linked to a series of binary oppositions, such as contamination/purity, savagery/civilization, authoritarian/democratic, and so forth. Within this timescape, people like Uncle San, and the larger group of Cambodians victims for whom he stands, are imagined in certain sorts of ways.

Uncle San lives within transitional justice time. The first line of the booklet delimits a temporal horizon of the past, as the introductory note explains, 'The Khmer Rouge Regime is generally recognized as the time between April 17, 1975 and January 6, 1979. This was a time in Cambodian history where the Communist Party of Kampuchea held control over the entire country and committed many crimes against the Cambodian people' (2).

Here, at the very start of the booklet, time is immediately constructed in three interlinked ways. First, in terms of periodicity, time is placed within a delimited period, 17 April 1979 to 6 January 1979, or the period of Khmer Rouge rule in Cambodia. This interval is then coloured in two ways. On the one hand, it constitutes a juridical frame, or what is called the temporal jurisdiction of the court. On the other hand, this interval is marked as one of criminality as a criminal act ('many crimes') has been committed by a perpetrator (the Communist Party of Kampuchea) against a victim ('against the Cambodian people'). There is no space in this temporal horizon for ambiguity; there are perpetrators and victims and nothing in between. Transitional justice time does not do well with 'grey zones'. This delimitation of time is further bound by a spatial framing, as the crimes take place in a particular national space, as opposed to a geopolitical space that is thereby erased, suggesting the problem was solely internal to Cambodia.

The booklet provides a quick overview of what happened in this spatiotemporal context through the eyes of Uncle San, a moustached Cambodian villager who wears a checked yellow Cambodian scarf slung over his right shoulder. 'Hello, my name is San,' his story begins. 'I am 64 years of age … I have lived in this village since I was young, but during the Khmer Rouge Regime I was forced to live in another area' (4). The accompanying graphic shows Uncle San sitting cross-legged on a table-like platform telling several of his fellow villagers about his forcible eviction. The importance of his experience is emphasized by the attentiveness of those gathered around, including two young children, as he tells his story.

Uncle San's experiences could be those of any Cambodian village survivor. Indeed, the Director of the KID told me that they selected the names Uncle San and Aunty Yan because they were 'common names among [the] rural population … very poor, grassroots type of names'.[7] Individual difference is thereby compressed as Uncle San's experiences could be those of any rural Cambodian survivor. The image asserts Uncle San's everyman status – he stands as an emblem of the Cambodian survivor-victim and, through metonymy, of Cambodia itself.

And he, and Cambodia by extension, exists in a deeply troubled state, one that suggests a lack and failure ('a failed state'). On page 5 of the booklet, Uncle San dozes fitfully in a hammock, dreaming about the Khmer Rouge past, which is represented by four recollections: his stealing a crab; and the Khmer Rouge torching a village, menacing two Buddhist monks who are being forcibly disrobed, and executing a man who looks like Uncle San in a mass grave filled with skulls. As in earlier shots, all of the characters resemble the moustached Uncle San. He is everyone and yet no one. He tells us:

> During the Khmer Rouge Regime, I was forced to plant rice all day long. Once, I took a crab from the field and was beaten for doing so. I remember the mistreatment of monks, hard work, poor food, tortures, and killings. When it was over and I went back to my village, my home was destroyed. What makes me most sad is that all of my family members were killed. Since then, I have bad dreams every night about what happened.
>
> (6)

In the next frame, Uncle San sits chatting with Aunty Yan and some other villagers in a rustic, traditional village space, one that lacks signs of modernity (for example, electricity, cars, motorcycles, industry, upscale commodities). 'Usually', Uncle San tells us in the accompanying text, 'I try not to think about the past by spending my time planting rice, going to the pagoda, and chatting with my neighbors.' Aunty Yan, it turns out, is a childhood friend who also lost her family during the Khmer Rouge regime and with whom Uncle San often shares meals or drinks tea.

If time is partly one of criminality in the booklet, it also suggests a pre-existing stasis. Perpetrators bear the impurity of their act, an unchanging stigma marking them as nefarious. Victims, in turn, remain wounded and unhealed, awaiting rescue. Thus Uncle San reveals, 'Since then, I have had bad dreams every night about what happened' and he tries 'not to think about the past' by keeping busy. This spatiotemporal freezing is indexed grammatically, as the temporal marker, 'since then' (*chap teang pi pel nuh mok*) frame the pronoun 'I' (*khnom*). Uncle San, and, as the everyman, by implication all Cambodian victims of the Khmer Rouge, lives suspended in a past of traumatic experience, which persists, unchanging, through a set of symptoms, including re-experiences (flashbacks, bad dreams, and nightmares), avoidance behaviours (trying 'not to think about the past'), and, as the frowning photo of him in the hammock suggests, hyperarousal (difficulty sleeping and feeling tense).[8]

Indeed, there is a direct relationship between transitional justice time and one of the subject positions produced by the transitional justice imaginary, that of 'the trauma victim'. We could even speak of a sort of pop-psychology

Figure 3.1 Uncle Sam's experiences in the booklet

'trauma imaginary' that overlaps with the transitional justice imaginary. In this trauma imaginary, individuals are viewed as existing in a state of regressive dysfunction, trapped by the seeds of past trauma. These 'seeds' manifest themselves through the aforementioned set of symptoms (the 'shoots' of the trauma 'seeds'). The trauma victim is more or less helpless until saved.

Within the transitional justice imaginary, the helpers are a legion of psychosocial specialists who treat trauma victims in post-conflict situation. But the transitional justice mechanism itself is depicted as constituting a form of treatment, evident in the frequent assertion that transitional justice mechanisms will help the society 'heal'. The post-conflict society is metonymically represented by the 'trauma victim', frozen in a regressive, impure, backward 'pre-' state until liberated by transitional justice and its practitioners, who launch it forward to a pure, progressive, liberal democratic state through a given form of 'treatment', the transitional justice mechanism (mirrored by the healing of the trauma victim through engagement with transitional justice practitioners and practices, especially the mechanisms in question).

On pages 9–10 of the booklet, this past and present of transitional justice time come sharply into focus. The left-side graphic shows a split scene of Uncle San lighting incense and praying for the spirits of the dead juxtaposed to an Aunty Yan speech bubble as she hails him to come to a 'special meeting about the Khmer Rouge … in our village' being conducted by a 'Citizen Advisor' from the Khmer Institute of Democracy (KID), the non-governmental organization that produced the booklet to use in its tribunal outreach activities.

Figure 3.2 Transitional Justice time in the booklet

Here we see transitional justice time in motion, as Uncle San moves from his frozen, backward states (performing ineffective 'traditional' practices incapable of healing his trauma) to an active, progressive, civilized states (engaging with the court, which has the potential to heal the traumatic wounds that have afflicted him for so long). In this context, civil society practitioners serve as the mediators of transformation as they bring the court to the villages that are so distant from it, a practice aptly termed, as noted above, 'outreach'. The court reaches out and provides its healing touch to even those living in the remote countryside with the help of organizations like KID. Particularly during the early phases of the court, which started operation in mid-2006, a number of these organizations served as intermediaries between the court and the population. Many of these intermediary organizations had been established during or soon after the 1993 United Nation-sponsored elections in Cambodia to promote human rights, law, and democracy. Over time, each developed particular areas of focus and distinct mechanisms to fulfil their missions.

KID, for example, was established on 6 October 1992, just prior to the UN-backed elections, by a group of Cambodian-Americans and Ambassador Julio Jeldres, an advisor to and official biographer of King Norodom Sihanouk.[9] KID's homepage states that the NGO's mission is 'to foster democratic values in Cambodian society by maintaining a neutral political position'. To this end, KID 'carries out a number of activities to promote a liberal democratic order as determined by the 1991 Paris Peace Agreement of 1991, based on a multi-party liberal democracy system, on human rights, and the respect of law as stipulated in the Constitution'.[10]

This mission is reflected in KID's Khmer Rouge Tribunal outreach programme, which began in November 2005 and involved tens of thousands of villagers in seven provinces. It was carried out with the goal of 'eliminating Cambodia's culture of impunity, ensuring respect for the rule of law, and facilitating people's participation in the tribunal process'.[11] KID's 184 'citizen advisors', comprising respected and more educated villagers such as teachers, stand at the heart of this and other KID initiatives. Inspired by the citizen advisors created in Britain during World War II, KID's citizen advisors seek to help and inform 'local people in remote areas, where there is a poor knowledge of democracy and limited respect for human rights' and thereby 'promote understanding of the law and its administration, how to prevent and resolve conflict, and how to promote peace in the community'.[12]

Starting in 2007, KID's citizen advisors were tasked with conducting outreach workshops with groups of 25–30 people in different villages.[13] The sessions provided a basic explanation of what the tribunal was and how it operated as well as how victims might become involved. Indeed, the citizen advisor also assisted those who were interested in filling out applications to become civil parties, complainants, or witnesses at the tribunal. Prior to the start of this initiative, KID's outreach team had begun to develop a variety of outreach materials, ranging from a flip chart and films to several explanatory booklets, including *Uncle San, Aunty Yan, and the Khmer Rouge Tribunal.*

In the speech bubble juxtaposed to the image of Uncle San praying to the spirits of his ancestors, an 'interested' and 'curious' Uncle San is hailed, like the reader of the booklet, to participate in a KID outreach session. Standing in front of a European Union and a ECCC poster affixed to a wooden beam, a KID citizen advisor reads from one of the KID outreach booklets (several of them also hold KID booklets) to 20 villages. Uncle San relates how, 'at the meeting, KID's CA described the Khmer Rouge Trials', about which the villagers had never before heard. He learns that the tribunal is located in the capital of Cambodia, Phnom Penh, and is composed of international and national staff.

The citizen advisor also tells Uncle San and his fellow villagers how the United Nations and the Cambodian government reached an agreement to hold the tribunal in 2003 in order to 'to seek justice, national reconciliation, stability, peace, and security in Cambodia' (p. 12). A graphic shows Cambodian and UN officials in suits signing the agreement. No one in this frame looks like Uncle San. Here we meet transitional justice time head on as we jump from the Khmer Rouge period (1975–79) to the origin of the transitional justice mechanism in 2003. Uncle San, the Cambodian everyman who is locked in a static, traumatized, primitive, and savage time of the past, steps into the progressive, healing, developed, and civilized time of the present transitional justice moment. What happened prior to 1975 and between 1979 and 2003 is flattened and erased,[14] a temporal erasure that is one of the hallmarks of dichotomous transitional justice time.

In other words, transitional justice time dehistoricizes in contrast to the truth claim with which it is often associated. We learn nothing of the origins or the immediate aftermaths of the conflict, such as the Vietnam War or the geopolitical politics that helped civil war until the late 1990s. After being toppled in early 1979, the former genocidaires were rearmed by the US, China, Thailand, and others – and even given Cambodia's seat at the United Nations. Uncle San does not occupy this time. There is just the Khmer Rouge period and then the 2003 agreement. Without it, there is no progression, just stasis. Uncle San, like the 'failed' state of Cambodia, is frozen in time until this moment, in which he is remade. Despite the oft-heard claims that the Khmer Rouge Tribunal will reveal 'the truth', transitional justice time involves erasures, as a broader understanding of Cambodian society, history, and geopolitics, factors that provide the critical backdrop and aftermath of the genocide, disappear from sight, diminishing understanding and producing an eclipsed truth.

In the next frame, Uncle San and Aunty Yan are seen sitting on a Cambodian platform bench in front of her cement house, which suggests somewhat greater wealth and status – and perhaps education – than Uncle San, who lives in a traditional wooden house. A radio hangs from a window. Uncle San holds a pen and looks expectantly toward her as they fill out forms. He states: 'Aunty Yan knows more about the KRT than me because she listens to the radio every day. Aunty Yan taught me that the victims can submit a complaint to the court. She showed me the complaint form and taught me how to fill it out' (14).

This scene suggests some of the key normative goods underlying the transitional justice imaginary, particularly freedom of choice and equality. One the one hand, Uncle San has the right and freedom to choose whether or not to participate in the tribunal. He asserts this liberal subjectivity by stating, 'I think I want to [become a complainant]!' On the other hand, the positions of Uncle San and Aunty Yan reverse traditional village gender norms, where the man would normally be assumed to speak from the position of authority.

The KID project officer who was in charge of developing the booklet told me that the picture was drawn this way because 'we have a human rights project. Here it is the influence of the human rights concept … of gender'. He stated that they wanted to combat the notion that 'women don't know anything in the grassroots [level]' and to teach people 'to not look down on the women in the community, but to show that every person has the same rights and dignity'. This focus on gender equality can also be seen in the legal proceedings, where gender-based crimes, such as sexual violence and forced marriages, have been foregrounded.

The page concludes with Uncle San stating, 'Aunty Yan and I also want to take a trip to the ECCC.' Most of the remainder of the booklet describes their journey and experiences there. Their mode of transportation is a sleek, modern bus with an ECCC logo on the side. The project officer explained that the bus was meant to reflect the 'international standards' of the court.

Here we find a spatiotemporal progression that mirrors the transformation of consciousness taking place. Uncle San's position of stasis is first destabilized by the KID village outreach programme, which 'hails' him toward the transitional justice imaginary. There, he begins to learn the outlines of this vision, manifest in discussions about the court and its operations. He is invited to become a part of the process, first by considering becoming a complainant ('I think I want to') and then, through the bus ride, directly entering into the spatiotemporal zone of the court, thereby 'leaving behind' the 'static' and 'less developed' village. All of this could readily be viewed through the anthropological lens of a rite of passage, in which Uncle San passes through a threshold, boarding the 'international standards' bus and crossing the gates of the court (behind which, in the booklet, stands the courtroom building, which is modern yet looks almost royal with towering Khmer spires in the background) and is transformed through ritual activities (legal procedure in the broadest sense) in this liminal space.

The remainder of the book describes Uncle San and Aunty Yan's trip to the Khmer Rouge Tribunal, where they learn about the court and Uncle San meets with the head of the Victims Unit, who tells him that if he fills out an application that is accepted by the judges 'based on several conditions of the law', he will become a civil party and 'have the right to participate in all of the court proceedings, plus a right to request collective and moral reparations' (24).

The day after their trip to the KRT, Uncle San and Aunty Yan are depicted discussing the tribunal with several of their fellow villagers. They have become like emissaries of the court – a manifestation of the longed-for 'multiplier effect' that one often hears mentioned in the outreach community – as they inform other villagers about what they saw and learned. The entry of this information into the consciousness of the other villagers is illustrated by a series of five thought bubbles in the graphic: a convicted criminal being led to jail, the scales of justice, the court itself, a well, and a stupa – images that contrast strongly with the images of Khmer Rouge violence that preoccupied Uncle San at the start of the booklet. 'We talked a long time about the KRT and the future of Cambodia', Uncle San tell us. 'We agreed that the establishment of the KRT is very good to seek justice for victims. The trials can find truth and give us relief (*sabay chhet*) from the past.' Here we find another manifestation of the normative dimension of the transitional justice imaginary: the notion that such mechanisms will deliver a set of goods such as truth, healing, moral reparation, and societal transformation. These normative goods thereby connected to the end point of the teleology driving transitional justice time toward a longed-for state of progress and development.

This seemingly simple booklet, so popular that apparently the court considered purchasing the rights to it, can be read in many ways. On the most obvious level, it provides an overview of the reasons for and structure of the court with a particular focus on victims' participation. In this sense it echoes, in a very general sense, much of the outreach message that the Khmer Rouge Tribunal and various intermediary organizations have been attempting to convey.

As such, it may also be read as a token of the court that symbolizes and condenses its larger meanings. Most broadly, I want to argue that the booklet is productive in two senses. First, the booklet embodies notions of transitional justice that are central to the larger functioning and legitimation of the court itself and are part of a larger transitional justice imaginary. This imaginary, as I have written elsewhere (Hinton 2010), suggests a teleology of a movement from a contaminated pre-state (of regressive savagery, violence, chaos, anarchy, etc.) to a purified post-state of a modern liberal democratic order (associated with what is civilized, peaceful, ordered, progressive, etc.), with the transitional justice mechanism – in this case the tribunal – serving as the mechanism of change.

This schema is directly manifest in the booklet. It begins with a coding of Cambodia as a place of violence, savagery, and regression, as Uncle San recalls the horrors of the Khmer Rouge. He himself embodies the regression, as he is plagued by dark memories of the past. He is a traumatized victim, childlike, an incomplete, not fully functioning being. Like Cambodia, he needs help to move forward.

The court is the vehicle of this transition. Indeed, the slogan of the court is 'Moving Forward through Justice'. The end state of the transition is stated in its basic goals: justice, reconciliation, peace, truth, and relief as we are told several times. The court itself signifies Cambodia's lack, manifesting the modernity it has not achieved. Uncle San notes the technology at the court even as the graphic images suggest the sleek, modern, high-tech nature of the court. Even the bus Uncle San and Aunty Yan take to the KRT must be of 'international standards'.

This mechanism already suggests the end, the post-state of modernity to be achieved. At the end of the process, Cambodia will attain what it lacks. Thus, after their visit to the court, Uncle San, Aunty Yan, and their neighbours discuss not just the court but the future of their country, the post-state. The accompanying graphics contains a picture of the court as part of a series of interlinked images that suggest this better future: criminals (who lived freely because of a 'culture of impunity' and a lack of 'the rule of law') are taken to jail as justice is upheld; the scales of justice balanced; a stupa symbolizing peace for the dead and reparations for the living; and a well signifying reparation, development, social justice, and repair. Once again, at the end, Uncle San himself embodies the new state of progress as he, like Cambodia, is healed and democratized by the process. He then sleeps through the night like a young child who has finally stepped forward into a blissful new stage of development.

And symbolically Uncle San is not the same. His very being has been transformed as he becomes (at least it is suggested) a modern liberal, rights-bearing subject who is healed through the process. Indeed, he now lives in a new world of modernity. The last graphic shows Uncle San sleeping in his hammock near a thatched house, where he dreams of a new Cambodia, like him remade, one with electricity, fancy wooden houses, and even a factory in the distance.

In contrast to the initial hammock frame in which a frowning Uncle San is plagued by the nightmares of the past, the last graphic shows him sleeping comfortably in his hammock, a slight smile on his face as he dreams of this new Cambodia. The accompanying caption reads, 'Then … I slept the whole night with no bad dreams' (34).

Uncle San, like Cambodia, is imagined as purified, renewed, and remade through the mechanism of the court as he passes through transitional justice time.

The KID project leader, who worked with an art student to design all of the graphics, was explicit about the message of this last frame: 'Here, Uncle San, after his participation, a long walk and journey, comes [back] to his own house. He can now close his eyes peacefully. He [dreams of] a peaceful situation and happiness.' Flipping back forth between the initial graphic of Uncle San's 'bad dreams' and his dream on the last page, he explained:

> This one [the first page] is tragedy, bad things, the [last page has] good things … birds and trees … kids who go to school in peace. The villagers have jobs [and] there is no mistreatment of monks … And you can see

Figure 3.3 Uncle Sam's new dream in the booklet

> [that the village] now has electricity … Normally only the rich have money to buy wood tile [houses]. So [this page] means that there is prosperity … no famine … [and] where we have factories, [we have] development.

The project leader explained that the meaning of the booklet was that people would live 'peacefully after participating in the court process. This is the real output we would like to explain to the grassroots … That is your benefit.' Uncle San, he continued, is a changed man, who no longer has psychological syndromes or bad dreams. 'We let the reader conclude that the court changed him because of his participation.' The factory, in turn, symbolized economic development in a rural landscape that normally lacks such industry. The people imagined in Uncle San's new dream bubble 'go to the factory to produce the final product [that is sold on the] market. That is the development process'.

Here the KID team leader explicitly describes the end point of transitional justice time: a liberal democratic order occupied by the functional, rights-bearing individual, capitalism, and, of course, the qualities that supposedly come with it: peace, happiness, and progress.

With Uncle San, the reader journeys through the transitional justice imaginary. Our minds, like his, become filled new thoughts and images, symbolically depicted by the thought bubble graphics. And like him, the reader symbolically passes through a transformative rite of passage and produces a new state of being. This imaginary asserts specific sorts of time (a transitional justice time characterized by temporal erasure, a teleology, and the instantiation of a series of pre- and post-state binaries), subjectivity (liberal, democratic, rights-bearing, juridical beings such as lawyers, civil parties, and even defendants whose fair-trial rights are frequently invoked), and moral economies of justice (the 'gift' of the international community and the sorts of normative goods it bestows, such as peace, reconciliation, healing, truth, justice). In this imaginary, even as the transitional society emerges, it achieves a still-fragile status of 'newly emerging democracy', one that is not on par with the implicitly 'mature' democratic governments and institutions – who are part of the 'international community' constitute in the transitional justice imaginary – that help guide the transition.

To seek to unpack the assumptions of transitional justice is not to simply dismiss it. It is to engage in a 'critical transitional justice studies' (Hinton 2010) that allows us to recognize the gaps within and shadows behind that which is assumed and naturalized. In particular, this imaginary has a tendency to erase historical and sociocultural complexities, ones that are directly relevant to the presumed normative goods of 'truth', 'prevention', and 'understanding the past' that are so often asserted in transitional justice rhetorics. Even for a strong supporter of transitional justice initiatives, such understanding is crucial, for it suggests alternative ways in which such mechanisms for dealing with the legacies of the past might unfold. To ignore such critical thinking is to risk remaining, like the initial construction of Uncle San, caught, unknowing, in the webs of the transitional justice imaginary.

Notes

1 *Special Note: Permission has been obtained to reproduce the images displayed in this chapter; it is understood that such reproductions will most likely be in black and white in the published book.

2 For more information about this period of Cambodian history, see: Chandler (1999); Hinton (2005); Kiernan (1996).

3 A 'supermajority' of judges is required for conviction, thereby ensuring that at least one foreign judge must join in any decision made by the Pre-Trial Chamber (four out of five judges), Trial Chamber (four out of five judges), or Supreme Court Chamber (five out of seven judges), each of which has a majority of Cambodia jurists.
4 On the origins and history of transitional justice, see Arthur (2009); Hinton (2010); Teitel (2003). On the components of transitional justice from a practitioner side, see the website of the International Center for Transitional Justice (http://ictj.org/about/transitional-justice, accessed 8 November 2011).
5 Alexander Laban Hinton, *The Transitional Justice Imaginary: Outreach at the Khmer Rouge Tribunal*. Newark, NJ: Rutgers University Press, forthcoming.
6 On the notion of social imaginaries, see: Anderson (2006); Castoriadis (1987); Taylor (2004).
7 Interview with Chhaya Hang, 23 June 2011.
8 On the symptoms of PTSD, see http://www.nimh.nih.gov/health/publications/post-traumatic-stress-disorder-ptsd/what-are-the-symptoms-of-ptsd.shtml, accessed 3 November 2011. The booklet seems to assume a Western biomedical model of post-traumatic stress disorder symptomology as opposed to local idioms of distress that are found in Cambodian villages (see Hinton et al., 2010).
9 The Khmer Institute of Democracy homepage, www3.online.com.kh/users/kid/index.htm, accessed 3 November 2011.
10 Ibid.
11 'Outreach Activities', Khmer Institute of Democracy, www.khmerrough.com/pdf/OutreachActivities.pdf, accessed 3 November 2011.
12 'Proto-Ombudsman Program (Citizen advisor Project)', Khmer Institute for Democracy, www3.online.com.kh/users/kid/program.htm, accessed 3 November 2011.
13 'Citizen Advisor Training and Outreach', Khmer Institute of Democracy, www.khmerrough.com/citizen.htm, accessed 3 November 2011.
14 For an overview of some of the events that took place during this time, see Fawthrop and Jarvis (2004).

References

Anderson, Benedict. 2006. *Imagined Communities: Reflections on the Origin and Spread of Nationalism*. New York: Verso.

Arthur, Paige. 2009. How 'Transitions' Reshaped Human Rights: A Conceptual History of Transitional Justice. *Human Rights Quarterly* 31(2): 321–367.

Castoriadis, Cornelius. 1987. *The Imaginary Institution of Society*. Cambridge, MA: MIT Press.

Chandler, David P. 1999. *Voices from S-21: Terror and History in Pol Pot's Secret Prison*. Berkeley: University of California Press.

Fawthrop, Tom, and Helen Jarvis. 2004. *Getting Away with Genocide: Cambodia's Long Struggle against the Khmer Rouge*. London: Pluto.

Hinton, Alexander Laban. 2005. *Why Did They Kill? Cambodia in the Shadow of Genocide*. Berkeley: University of California Press.

Hinton, Alexander Laban (ed.). 2010. *Transitional Justice: Global Mechanisms and Local Realities after Genocide and Mass Violence*. Piscataway, NJ: Rutgers University Press.

Hinton, Alexander Laban. 2010. Introduction: Toward an Anthropology of Transitional Justice. In Alexander Laban Hinton (ed.), *Transitional Justice: Global Mechanisms and*

Local Realities after Genocide and Mass Atrocity, pp. 1–22. Piscataway, NJ: Rutgers University Press.

Hinton, Devon, Alexander Laban Hinton, Kok-Thay Eng and Sophearith Choung. 2010. PTSD Severity and Key Idioms of Distress among Rural Cambodians: The Results of a Needs Assessment. In *Cambodia's Hidden Scars: Trauma Psychology in the Wake of the Khmer Rouge* (an edited volume on Cambodia's mental health), pp. 47–68. Phnom Penh: Documentation Center of Cambodia.

Khmer Institute of Democracy. 2008. *Uncle San, Aunty Yan, and the KRT.* Khmer Institute of Democracy, Phnom Penh, Cambodia.

Kiernan, Ben. 1996. *The Pol Pot Regime: Race, Power, and Genocide in Cambodia under the Khmer Rouge, 1975–79.* New Haven, CT: Yale University Press.

Taylor, Charles. 2004. *Modern Social Imaginaries.* Durham, NC: Duke University Press.

Teitel, Ruti G. 2003. Transitional Justice Genealogy. *Harvard Human Rights Journal* 16: 69–94.

4 Peace processes and social acceleration

The case of Colombia

Sandra Rios Oyola

Introduction

This chapter seeks to contribute to the study of time and temporality in post-conflict and transitional societies by highlighting the process of social acceleration at play during peace processes, through the lenses of the peace process in Colombia. In 2016, after four years of negotiations between the Colombian government and the FARC-EP guerrillas (Fuerzas Armadas Revolucionarias de Colombia – Ejército del Pueblo – Revolutionary Armed Forces of Colombia – The People's Army), a peace agenda was agreed. As part of this process, the parties decided to call for a plebiscite in order to legitimise the agreement by popular vote. The months preceding the plebiscite were filled with strong campaigns supporting and opposing the peace agreement; additionally, the FARC performed a number of public apologies to the victims of some of their most egregious crimes. In a surprising outcome, the opposition to the peace process won the plebiscite by a very small margin.

The campaigns that preceded the peace plebiscite were the battleground not only for different means to resolve the conflict but for the imposition of different temporalities. Following the literature on social acceleration, this chapter analyses the peace plebiscite and the FARC's public apologies as devices that sought to reduce the de-synchronisation between the political peace process and the social peace process. The chapter reviews the arguments for understanding a peace process as a process of social acceleration. The chapter then provides a brief overview of the Colombian peace process in order to situate the peace plebiscite and the FARC apologies in context. The content of one of the apologies presented by the FARC as well as the peace plebiscite will be analysed.

A political peace process depends upon the actors involved at the negotiating table, while the social peace process is built by civil society which deals with the legacy of the conflict and attempts to rebuild the broken social fabric. Brewer (2013: 162) defines the latter as 'the process of societal healing, that is, the restoration of broken relationships, the development of a sense of community and shared responsibility for the future. Reconciliation is part of this, but so too is compromise, empathy, trust and forgiveness'. The political accord and the social peace process can influence each other. For instance, the negotiation at the political level opens the gate for the social peace process by facilitating the possibility of dialogue with the antagonist; in a different scenario, the social peace process starts as initiatives at the grassroots at a local level, providing support for the negotiations to come.

In Colombia, before the current peace negotiation with the FARC, there were several attempts to create a political peace accord between illegal armed actors and the government. Some of them were successful in reintegrating a number of the smaller guerrillas to civil society; others have reached mixed results, as in the case of the process of the demobilisation of the paramilitary, which officially ceased to exist in 2006, although new post-demobilised groups emerged (Nussio and Howe 2016). Simultaneously, there have been thousands of peace initiatives at the local, regional and national level; they include women's, indigenous, Afro-Colombian and peasant groups with peace agendas that respond to the different humanitarian crisis that have existed during the long history of the conflict (Bouvier 2009). Although the political and the social peace process are deeply intertwined, they respond to different temporalities. To analyse the relationship between the Colombian peace process at a political level and at a social level from the perspective of time and temporality means to identify how one type of peace process has influenced the other, including the strategies used to overcome the effects of the de-synchronisation between both spheres.

The chapter is based on qualitative research methods including fieldwork, participant observation and interviews in Colombia, particularly in the region of Bojayá, in 2012 and in 2016. Interviews were undertaken with social and victim leaders, religious actors and members of the national Historical Memory Group. Additionally, the news coverage of the peace process and peace plebiscite was analysed.

Time and peace processes

The study of the time and temporality of peace processes has often focused on debates around the ripeness of a conflict (Zartman 2000; 2001) and on sequencing of peacebuilding interventions (Langer and Brown 2016). The first debate follows the assumption that there is a right moment in a conflict when engaging in peace negotiations seems like a reasonable next step; this occurs when both parties have arrived at a stalemate position that does not provide a way out on the horizon other than engaging in a negotiated solution (see also Rosoux's Chapter 2 in this volume). The second debate considers a sequencing perspective, which is based on the coordinated efforts that should be prioritised in a peacebuilding context, such as socioeconomic development over political reconciliation, in order to make it successful and stable (Walton 2016). Although these debates are important for understanding the policy implications of different peacebuilding strategies, including what type of programmes and interventions are '*needed* at a particular *moment in time*'

(Langer and Brown 2016: 4), they do not expand our knowledge on the temporalities that dominate the political and social peace process respectively.

The political peace process is framed by the narrative of teleological movement between conflict and post-conflict state (Hinton, this volume). Political peace processes respond to the language of transition politics, which states that 'as societies move "forward", violence is "left" and "locked" behind' (Castillejo-Cuéllar 2014: 62). Political peace accords often include transitional justice mechanisms such as amnesties, trials, truth commissions or reparations. For instance, in Northern Ireland, the Good Friday Agreement established the provision of services and support to victims in order to recognise the harms that they suffered. Although the implementation of mechanisms of reparation is slow, these are principles that are quickly established in the peace agreement to move forward to the post-conflict stage. In the context of peacebuilding processes, Mac Ginty (2016: 15) distinguishes political time from sociological time. While political time 'may include dates for elections, the timing of ceasefires, or deadlines for peace negotiation, sociological time refers to non-elite concepts of time that may revolve around everyday activities of family life, work, and cultural pursuits'.

The social peace process is framed by the lived experience of conflict, which can create different temporal demands in civil society. For instance, long-term grievances resulting from structural forms of inequality and exclusion can affect subjective and local temporalities (Castillejo-Cuéllar 2014). The temporality of the social peace process might be slower than the temporality of the negotiating table, which responds to deadlines imposed by international collaborators, national pressures subdued to local and national elections, as well as the risk to damage a vulnerable peace agreement when visible reforms do not occur. The process by which the temporality of political peace processes tries to influence the temporality of social peace processes can be understood in terms of social acceleration.

Multiple temporalities and social acceleration

Hartmut Rosa (2015) is one of the main scholars of social acceleration; he argues that modernity follows a process of social acceleration that permeates all spheres of social life. Rosa explains that the important social changes that have taken place in late modernity have been fuelled by rapid technological transformation, which directly affects our pace of life and in turn demands new and faster technological solutions. This process is embedded in an economic logic of capitalism and generates the dominant cultural programme of modernity marked by acceleration; according to this logic, 'whoever becomes infinitely fast approaches the potentially unlimited horizon of the time of the world (and of worldly possibilities) within one lifetime to the extent that she can realise a plurality of life possibilities in a single lifespan' (Rosa 2015: 311). This culture has also affected the political project of democracy because the rapid succession of governments would reflect the rapid changes in 'the political sensitivity towards the evolving needs of society' (Rosa 2005: 449). The speedy changes that dominate political competition and the renewed political elections would reflect the debates that are relevant for that specific moment in society.

An important consequence of the process of social acceleration in social life is that certain processes often move faster than other spheres of social life; for instance, the technological development and the socioeconomic sphere moves faster than nature's tempo, creating the current ecological crisis. In the political sphere, socioeconomic transformations could move too fast for the legal debates to take place. Further, Rosa claims that the time, stability and centrality required for debating, deliberating and achieving collective will-formation 'might not be compatible with the most dynamic forms of society' (Rosa 2005: 450). Similarly, scholars such as Scheuerman (2004) and Hassan (2009) have criticised the high speed of current political institutions, which have evolved in response to neoliberal pressures to the detriment of the slowness required for the debate of ideas, face-to-face meetings and the working through arguments and recommendations. This type of neoliberal acceleration of democracy can lead to short-termism and to taking shortcuts that could affect the democratic process. Following Rosa, the de-synchronisation between different spheres of social life can lead to a crisis. In the case of societies in transition, this chapter argues that there is a de-synchronisation between the transformation that occur in the political sphere of the peace process and the social sphere of the peace process.

Societies in transition experience a unique moment of social transformation, where the *tempo* of the socioeconomic, political and legal institutions increases because it is subject to rapid socio-political transformations. There are many transformations that transitional societies go through. For example, illegal armed actors should be reintegrated to civil society; legal tools should be created and implemented in order to make amnesties, trials and reparations possible; and security forces should become depoliticised and respectful of human rights. Usually these changes are subject to tight deadlines due to pressures by the international community. Additionally, in the case of societies that have undergone peace processes as a path towards their transition, time is a key element because processes are vulnerable. This means that serious commitments to implement the agreements of the peace agenda should be visible otherwise the whole peace process could fail.

By contrast, societies in conflict are *slowed down* by the rhythms of the conflict: there is a quasi-ritualistic behaviour related to violent massive practices, repetition of atrocities, and a polarised political scenario where labels of perpetrators/victims abound and impede the transformation and mobilisation of society. The presence of warlords with invested interests in the continuation of the war, the permanent creation of victims who sustain renewed feelings of revenge that could open the gate for new cycles of violence, and the financial sustainability of the armed groups through drug trafficking, extortion and illegal trade, are reasons for the conflict to continue. Conversely, a peace process becomes fuel for the acceleration of societies, where the socioeconomic and cultural arenas need to catch up with the evolving legal and political arena. A peace process is an opportunity to reinsert the political culture into the accelerated discourse of democracy, but while this is a process that occurs at the level of the negotiating table or the political peace process, its *temporality* might be different at the local level of the social peace process.

In what follows, this chapter explains how the peace process in Colombia became the fuel for social acceleration and it analyses some of the mechanisms used in combating the de-synchronisation between the political and the social peace process. It argues that the peace plebiscite was a device that could fuel the acceleration of the political sphere. Similarly, apologies were a tool for accelerating the process of reconciliation and bringing it up to speed with the official political discourse.

The conflict and peace process in Colombia

One of the defining characteristics of the Colombian conflict is its length; with over fifty years it is considered the longest war in the Americas (Casey 2016). There has been a succession of several violent periods that can be tracked back to the Spanish conquest, followed by the wars of independence and the multiple civil wars that gave birth to the modern Colombian state. The current conflict started in the late 1950s with the creation of communist guerrillas and extreme right-wing illegal paramilitary groups. It should be noted that 'guerrilla warfare is about a lengthy process asymmetric fight, which includes attack-and-fight actions that seek to stretch state's capabilities' (Zarama 2015: 497). There are several actors involved in the conflict, including the FARC-EP guerrillas, the AUC (Autodefensas Unidas de Colombia/ United Self-Defence of Colombia) and the ELN (Ejercito de Liberacion Nacional/National Liberation Army), but this chapter will only focus on the FARC and the AUC.

As the largest and longest-established insurgency in Latin American history, the FARC's origins are often located in the leftist peasant associations excluded by the elites from the political pact that resolved the 1950s civil war. These feelings of exclusion and inequality were accompanied by an ill distribution of land. During the 1980s, many of the smaller guerrillas accepted amnesties and processes of reintegration into civil society. During Betancour's administration (1984–1988), the FARC also agreed to a ceasefire and a civil political arm, the Patriotic Union, was created with popular support. However, soon after that, over 5,000 electoral activists from this party were killed. The FARC amplified their power in the years to come, and their earnings increased thanks to drug trafficking-related activities such as transport and tax for safe passage (Collier and Hoeffler 2000).

The paramilitary forces can be traced back to 1965, when the Congress approved the laws allowing the legitimate self-defence of their citizens. Although these laws were suspended in 1989, illegal paramilitary groups under the auspices of the AUC expanded their structure considerably between 1997 and 2004. They were involved in a high number of human rights violations and, contrary to the insurgent FARC, were a pro-systemic criminal organisation that often worked in alliance with the official security forces. The AUC had a clear political plan and sought to penetrate the electoral arena. They also benefited from drug traffic and from the political and economic support from big property owners, agri-business and extensive livestock farmers (Hristov 2014).

In 2005, the paramilitary groups initiated their demobilisation after negotiations with former President Alvaro Uribe. Several transitional justice mechanisms were created, under the framework of the Justice and Peace Law (2005). They included alternative reduced penalties (between five and eight years) as a consequence of the beneficiary's contribution to peace, his/her cooperation with justice, fact-finding and the reparation of the victims. As a result, 30,000 ex-paramilitaries reintegrated to civil life. This process was, however, heavily criticised for its weak execution of reparations and truth recovery as well as the increasing rates of impunity associated with it (Saffon and Uprimny 2009).

The rural regions of Colombia, which are more distant from the capital centre, have been particularly affected by the violence. There have been 2,087 massacres from 1983 to 2012. In addition, 5.7 million people have been displaced by force and 220,000 civilians have been murdered; 25,000 have been victims of forced disappearance; 1,754 have been victims of sexual violence; and 5,712,506 have been victims of forced displacement (Comisión Nacional de Reparación y Reconciliación (Colombia) 2013). Despite these unfavourable circumstances, during this period of time diverse and numerous peacebuilding initiatives were created among the grassroots communities across the country.

Juan Manuel Santos Calderon, who was previously Minister of Defence during Uribe's administration, became president in 2010. His administration took a different turn from Uribe's government and recognised the existence of a political armed conflict in Colombia instead of a war against narco-terrorist guerrillas. In 2011 the Law of Victims and Land Restitution was enacted, which created a number of mechanisms to repair victims and support the restitution of their land property. In June 2012, the Congress approved the Framework Law for Peace and the peace negotiations between the government and the FARC guerrilla started in September in Havana, before agreeing on a ceasefire.

Although the negotiations were done in Havana under controlled sharing of information, it was a relatively transparent process in which several commissions were invited, including victims' groups and women's groups, and the different topics were discussed in open forums in Colombia and later brought back to Havana. The parties agreed on a peace agenda in 2016 and, in October, they called for a plebiscite in order to ratify the deal. The negotiations involved several points that intended to address not only the process of demobilisation, weapon surrender, and reintegration of the demobilised guerrilla members, but also sought to address the root causes of the conflict, such as integral rural development, political participation and the problem of illicit drugs and victims.[1] The implementation required a rapid transformation of the legal framework, for which the negotiating committee and the president called for the creation of a 'fast-track' option. This channel would stimulate the fast discussion and approval of the laws required to implement the peace accord while protecting it against the changes that the Congress could try to introduce later and violate the accord's commitments. At the time of writing this chapter, the fast-track channel continues to be challenged in the Constitutional Court, and what was considered to be a rapid path towards the transformation of the legal system is currently being slowed down, particularly due to interests motivated by the upcoming elections (Semana 2017).

The FARC's apologies

During the last months of the peace negotiations, the FARC apologised for the harms that they occasioned to the victims. They carried three emblematic acts of apology: to the victims of the Bojayá massacre; to the relatives of the twelve Valle del Cauca's Departmental Assembly deputies, who were kidnapped and murdered in 2002; and to the victims of the Chinita massacre. This type of apology can be located in the framework of what some authors have called political forgiveness or political apology. According to Coicaud (2009: 95) a public apology involves 'acknowledging a wrong, admitting guilt, taking responsibility, recognising somebody's inflicted suffering, but also seeking to reverse victimisation, re-establish trust, empower the powerless, and end cycles of resentment and retribution'. It is considered to be a device to recognise and try to amend past wrongs.

Political apologies have been studied as a means to restore and re-define the political community and as a tool to bring the past to a closes. Further, Bastian (2014: 96) argues that 'official apologies are prime examples of attempts to intervene into shared understandings of political community and its temporality'. In her research on the public apologies performed by the Australian government regarding its role in causing the forced removal of Aboriginal children from their families from 1910 into the 1970s, she concludes that apologies pursue the goal of unifying time as an aspiration of the unified community. The apology implies a desire for 'putting the past to rest', and in doing so, it 'also works as a device for dividing time in order to separate different experiences of community from each other and so re-time the nation' (Bastian 2014: 104).

Contrary to Bastian's research, the political apologies studied in this chapter do not come from the government or the official establishment, although they are framed by the official peace negotiations. In this case the apology is less 'a struggle for which temporality will dominate' and more an effort to accelerate the time of the social peace process and bring it into synchrony with the political peace process. Moreover, this chapter does not only analyse the role of the political apology but also examines the response performed by the victims.

While perpetrators perform political apologies, victims are expected to forgive them; however, it is important to clarify that political forgiveness does not constitute individual forgiveness writ large. This means that political forgiveness is analogous to the relationship between individual and collective memory; it goes beyond the individual and encompasses power relations, interests, emotions and dynamics that are not necessarily part of individual forgiveness (Griswold 2007). According to Misztal (2016), political forgiveness is different from private interpersonal forgiveness because it occurs in the public sphere, where traumatic memories are retrieved: 'it depends on each party's truth telling in both an allocation of blame and an account of injury' (Misztal 2016: 5). Another important characteristic of political forgiveness is that is essentially conditional. In the public sphere of democratic regimes or regimes that aspire to be democratic, forgiveness should be conditional, neither expected nor pre-defined. Following Ricoeur's idea that even conditional forgiveness can repair fractures (Ricoeur 2004, 458), Misztal (2016) argues that forgiveness should be conditional because victims should have the choice to demand the conditions in which it would be possible to forgive. In what follows, I will focus on one of the above-mentioned acts of apology, the one the FARC presented to the Bojayá massacre victims and their response in terms of political forgiveness. The reason for the selection of this case is that it was one of the most emblematic crimes perpetrated by the FARC.

Forgiveness in Bojayá

In 2002, during a confrontation in the municipality of Bellavista in the Pacific lowlands of Colombia in Bojayá, the FARC guerrillas launched several homemade mortars assembled with gas cylinder parts (known as *pipeta* in Spanish) against the AUC paramilitary army, who were using the church San Pablo Apostol as a shield. In the church, over 300 Afro-Colombian civilians were seeking protection. The *pipeta* reached the church instead of the paramilitary group and as a result a non-determined number of victims were killed; although the community claims that 119 victims perished there (GMH-CNRR 2010).

Despite the large number of massacres that have occurred in the country, the victims of Bojayá have become an emblematic group of the FARC's victims during the past fifteen years (Rios Oyola 2015, 95). They represent the 'innocent victim', children, women and the elderly who were trapped in the crossfire. They also represent the chronic condition of government's abandonment the structural violence resulting from poverty and the struggle for being recognised as equal citizens (Gomez 2016). In the context of the peace negotiation between the FARC guerrillas and the Colombian government (2012–2016), the image of the FARC's most emblematic victims was used by deniers and supporters of the peace process. The opposition used the image of the Bojayá victims as a reminder of the FARC's atrocities while the supporters highlighted the victims' willingness to forgive as evidence that 'moving forward' was a possibility. They were seen as 'moral beacons' (Brewer and Hayes 2011). In this context, the FARC's public apology and (political) forgiveness by Bojayá's victims became relevant for national debates.

Several groups of victims' representatives were invited to the negotiation table in Havana, among them the victims from Bojayá. During the visit in December 2014, the FARC apologised for the harms they had caused in Bojayá and promised to take steps towards their reparation:

> Since then, this event has caused pain in our guerrilla soul, and that is the reason why we should express our sorrow at this moment, when we are dialoguing in order to build peace with truth and justice, as we expressed days after the event; we feel deep sorrow, the fatal outcome of this terrible event for the people from Bojayá hurts us deeply.
>
> (FARC member quoted by El Espectador 2014)

Although the event ended with hugs between victims and perpetrators, the representative leaders of the victims argued that 'only each victim can offer forgiveness, each survivor, that is why we are taking the FARC's declaration of apology request to our communities. It is a necessary step for starting an authentic reconciliation in the country to hear an apology from all the actors' (Delegación de Bojayá 2014).

After this initial conversation, victim leaders discussed the possibility of allowing the public act of apology with members of the community of Bellavista and neighbour communities in Bojayá (Fundación Comisión de Conciliación Nacional / Óscar F. Acevedo 2016, 185). During the preparatory steps for the FARC's public apology, there were intensive debates among the local communities about the possibility of forgiveness. The issue of interpersonal vs. political forgiveness was discussed in the community. They considered that forgiveness was an individual decision but allowing the FARC to conduct their public apology was a collective one. The multiple debates in the local communities led to the favouring of the FARC's public apology but under certain conditions, such as the absence of the media during the event.

The discussion in the local communities before the FARC's public apology provided the opportunity for collectively constructing a shared appraisal of the situation, and a realisation of overlapping private and collective concerns about the situation. According to Delys Palacios, a leader of the Committee for the Victims' Rights of Bojayá: 'I am not very hopeful about [the collective reparation] but I see it as an opportunity, this is a political opportunity to make Bojayá visible, to show the resources that we have.'[2]

Despite the debates and workshops in advance to the event, there was tension in the moments before the public apology by the FARC; for instance, a young victim reported that 'the thought of forgiving is there but once one is in front of [the perpetrators] one does not know what could happen' (Rivera 2015).

The *act of recognition of responsibility and public apology by the FARC* took place a year after the FARC's request, on 6 December 2015 in Bellavista. The Victims' Unit, the Historic Memory Center, the Catholic Church, the Peace commissioners, were also present during the event that was attended by 700 victims. During the public apology, Ivan Marquez, FARC leader, said:

> We know that these words, as we have expressed it on several occasions, do not repair that which is unrepairable, neither do they return any of the people taken away; they do not erase the enacted suffering. This suffering is expressed in the faces of every one of you, who we hope one day can forgive us.
>
> (Annex 2 in Gomez 2016, 330)

Similarly, Sergio Jaramillo, the governmental peace commissioner, expressed that 'this is the symbolical place to turn the page of the war … the essence of the change is to recognise what happened and accept the responsibility of what happened'.[3] Finally, the victims' associations presented their account of the events and the consequences of the presence of armed actors in their territory, emphasising the long chain of abuses that have harmed their communities, such as the lack of freedom of movement across the river and the forced emplacement of some indigenous communities. They asked for mechanisms of recognition and reparation for victims and collective reparation for the communities of the region and seem hopeful towards the possibility of reconciliation:

> Our boats have carried trust, which is the basis for starting a process of reconciliation, that is why we want to believe that this recognition of responsibility is based on Truth, in order to achieve Justice and Reparation for all our communities. May God want that soon other actors' recognition of responsibility of our tragedy be open.
>
> (Annex 1 in: Gomez 2016: 320)

The main condition for forgiveness was the cessation of violence. This corresponds to findings from my previous fieldwork in 2012, when the issue of forgiveness was not present, mainly because of the ongoing conflict. Victims perceived the act of political apology by the FARC as an act of public recognition of their responsibility in the events that took place in 2002. They had asked the FARC members, in advance of the public event, that they should consider the enactment of measures that could prevent or deter issues such as recruiting civilians, clearing land mines, extortion and illicit crops (Restrepo 2016). The year after the public apology and when some of the promises of reparation should have been in operation, the plebiscite for the peace process took place.

Process not moment (*Proceso no coyuntura*)

The conditions established by the victims' association in order to accept the FARC's apology reflect their interest in being recognised not only as victims of a specific atrocity, *the* massacre. Instead, they wanted recognition for the harms that they have suffered during the chain of events that have taken place in the region since colonial times but, specifically, in the last decades that they had had to struggle for the recognition of their rights over the territory. According to Gonzalo Sanchez, director of the Historical Memory Group in Colombia:

> The Bojayá community taught the country and to this peace process that there are tasks that are of long duration, in their creation and in their continuity. In acts like this [public apology], not only the insurgent groups (in this case), or whoever would do it in other cases, should answer to the victims' demands, but reparation and reconciliation are only possible if there is an effective answer to their demands and not only an instrumentalisation of their expectations.
>
> (Sanchez 2015)

Victims have used the commemoration and the FARC's public apology not only as a means of reconciliation and societal healing but as a platform where the responsibilities for the multiple forms of violence they have suffered should be recognised. This involves the role of other actors, such as the paramilitary and the state who did not respond to the early warnings issued by the local authorities and leaders. These claims go beyond the event of the massacre in 2002 and intend to cover the systematic attack against the black and indigenous communities, which could be considered a crime against humanity (Rios Oyola 2015: 112–117).

The peace plebiscite and the acceleration of forgiveness

Referendums or plebiscites can be suitable devices for conflict resolution, as they have the potential of locking the public opinion, creating coalitions and justifying powerful international support. A positive peace referendum can force the opposition to use alternative strategies to violence in order to gain support. Lee and Mac Ginty argue that peace referendums can be successful only when 'they are conducted at a specific moment' (Lee and Mac Ginty 2012: 45). However, the moment chosen for the Colombian plebiscite was affected by the polarisation that the country was experiencing. The plebiscite included one question, whether Colombian citizens would agree or not to the peace agreement reached by the government and the FARC. The plebiscite took place on 2 October 2016 and the campaign in preparation of the plebiscite lasted a few months. During the campaign against the peace process, severe attempts to misinform the public were reached, as was revealed by the very leader of the 'No' campaign, former senator Juan Carlos Velez (Ramirez 2016). The plebiscite opened the gate for a systematic and coordinated attack against the peace process.

Former president and current senator Alvaro Uribe skilfully presented a rhetoric of distrust and retribution against a possible peace process, diminishing the FARC's acts of public apology as not real acts of remorse (Blu Radio 2016). The opposition presented the peace process as an opportunity of forgiveness that the majority was not willing to negotiate. In the meantime, the supporters of the peace process reinforced a language of hope that presented the possibility of new futures that could allow escaping from new cycles of violence. Forgiveness was presented as the gate for opening the door to new possibilities, which could allow bringing together for the first time, guerrilla members army members and victims.

The plebiscite intended to legitimise the agreement reached between the government and the FARC but contrary to that desired effect, the plebiscite increased the politicisation of the peace process. This means the temporality of the political peace process was intended to be applied to the temporality of the social peace process, in an attempt to accelerate the processes of reconciliation, trust building and repair of the social fabric. Although the social peace process had been working even before the political process had started, its main area of influence was located in the periphery, in the regions most affected by the violence, which voted 'yes' to the plebiscite. Bojayá was one of these regions, where 96% of the voters supported the peace plebiscite (Semana 2016). However, in the cities, the social peace process was being slowed down by the opposition. During heated opposition campaigns, the issue of political forgiveness was questioned. The opposition saw in the peace agreement a pathway towards political forgiveness of the FARC, which many of them were not willing to concede yet.

The language of forgiveness changed during the campaign – for instance, the local Catholic Church, which had worked with the communities in the region for the last forty years – had until that moment accompanied the process of public apology without major emphasis on forgiving. During the peace campaign the language of collective forgiveness became stronger; the need for forgiveness was stressed in the use of messages such as the 'prodigal son' that is welcomed back in the home (Diocese of Quibdó, homily, September 2016) intended to favour a vote for the peace process. At an institutional level, the Catholic Church presented itself as neutral after a series of failed attempts to influence the public opinion that had resulted in the estrangement of a sector of the religious membership. Nevertheless, the Church actively participated as local and national facilitators and supported activities of information. The cathedral of Quibdó in Chocó, the department where Bojayá is located, hosted a series of activities in support of the peace process.

During the campaign, the open and reflective exercise of forgiveness became more politicised and the pressure of presenting the Bojayá's victims as willing supporters of the peace process increased. In online social networks and in the news, the memory of Bojayá was presented as a rhetorical device, a reason for not forgiving the FARC. Meanwhile the main counter-argument of the peace supporters was that Bojayá's victims were willing to forgive the FARC, therefore the rest of society should or could also do the same. The careful balance between individual and collective forgiveness that was worked through during the workshops and local debates was disrupted because of the acceleration fuelled by the peace plebiscite.

Conclusion

After the failure of the peace plebiscite, President Juan Manuel Santos was awarded the Nobel Peace prize for his efforts for bringing peace to the country. This international support helped to overcome the domestic failure and the peace agreement was ratified through the Congress. This international support powered the domestic support to the peace agreement, which was visible through numerous demonstrations on the streets.

This chapter has shown that there are not only different temporalities between the political and the social peace process, but that the temporality of reconciliation, trust and forgiveness is heterogeneous among civil society. In the Colombian case there were thousands of peace initiatives in place before the political peace process had started. There were active supporters of the peace plebiscite and activists even after its defeat. The peace plebiscite and the FARC's apologies were instruments to accelerate the social peace process; however, they had to compete with alternative versions of temporality defended by the opposition. This group sought to represent the grievances against the FARC and a military solution to the conflict. Peace spoilers slow down the rapid legal and social transformations that are necessary to build peace.

Notes

1 Documents pertaining to Colombia's peace dialogues in Havana can be found at www.mesadeconversaciones.com.co.
2 Reparación colectiva en Bojayá. VII asamblea Bellavista, 4–7 agosto 2016. https://soundcloud.com/user-894967417.
3 Sergio Jaramillo Acto de reconocimiento de responsabilidad FARC-EP en Bojayá, www.youtube.com/watch?v=IolyEHM2pL8 (accessed 26 March 2017).

References

Bastian, Michelle. 2014. Political Apologies and the Question of a 'Shared Time' in the Australian Context. *Theory, Culture & Society* 30(5): 94–121.

Blu Radio. 2016. Farc Ha Pedido Perdón Sin Arrepentimiento: Álvaro Uribe. *BLU Radio*. 30 September. www.bluradio.com/barranquilla/farc-ha-pedido-perdon-sin-arrepentimiento-alvaro-uribe-118051.

Bouvier, Virginia Marie (ed.). 2009. *Colombia: Building Peace in a Time of War*. Washington, DC: United States Institute of Peace.

Brewer, John. 2013. Sociology and Peacebuilding. In Roger MacGinty (ed.), *Routledge Handbook of Peacebuilding*, 159–170. London: Routledge.

Brewer, John, and Bernadette Hayes. 2011. Victims as Moral Beacons: Victims and Perpetrators in Northern Ireland. *Contemporary Social Science* 6(1): 73–88.

Casey, Nicholas. 2016. Colombia and FARC Reach Deal to End the Americas' Longest War. *The New York Times*, 24 August, sec. Americas. www.nytimes.com/2016/08/25/world/americas/colombia-farc-peace-deal.html.

Castillejo-Cuéllar, Alejandro. 2014. Historical Injuries, Temporality and the Law: Articulations of a Violent Past in Two Transitional Scenarios. *Law and Critique* 25(1): 47–66. doi:10.1007/s10978-013-9127-z.

Coicaud, Jean-Marc. 2009. Apology: A Small yet Important Part of Justice. *Japanese Journal of Political Science* 10(01): 93. doi:10.1017/S1468109908003393.

Collier, Paul and Anke Hoeffler. 2000. Greed and Grievance in Civil War. SSRN Scholarly Paper ID 630727. Rochester, NY: Social Science Research Network. https://papers.ssrn.com/abstract=630727.

Comisión Nacional de Reparación y Reconciliación (Colombia), (ed.) 2013. *¡Basta Ya! Colombia, Memorias de Guerra Y Dignidad: Informe General.* Segunda edición corregida. Bogotá: Centro Nacional de Memoria Histórica.

Delegacióonde Bojayá. 2014. Verdad, Justicia Y Reparación Para Las Comunidades de Bojayá-Chocó. *Bojayá Una Década.* 18 December. https://bojayaunadecada.org/tag/negociacion-habana/.

El Espectador. 2014. Bojayá, El Perdón Que Ofrecen Las Farc. Text. *EL Espectador.com* 18 December. www.elespectador.com/noticias/paz/bojaya-el-perdon-ofrecen-farc-articulo-533985.

Fundación Comisión de Conciliación Nacional / Óscar F. Acevedo. 2016. El corazón de las víctimas: Aportes a la verdad para la reconciliación en Colombia. Editorial San Pablo.

GMH-CNRR. 2010. *Bojayá: La Guerra Sin Limites.* Bogotá: Ediciones Semana, Taurus.

Gomez, Paco. 2016. *La Guerra No Es Un Relampago. Bojaya Hbala de La Guerra Y La Paz En Colombia.* Bogotá: Icono.

Griswold, Charles. 2007. *Forgiveness: A Philosophical Exploration.* Cambridge: Cambridge University Press.

Hassan, Robert. 2009. *Empires of Speed: Time and the Acceleration of Politics and Society.* Leiden and Boston, MA: Brill.

Hristov, Jasmin. 2014. *Paramilitarism and Neoliberalism: Violent Systems of Capital Accumulation in Colombia and beyond.* London: Pluto Press.

Langer, Arnim, and Graham K. Brown, (eds). 2016. *Building Sustainable Peace: Timing and Sequencing of Post-Conflict Reconstruction and Peacebuilding.* Oxford: Oxford University Press.

Lee, Sung Yong, and R. Mac Ginty. 2012. Context and Postconflict Referendums. *Nationalism and Ethnic Politics* 18(1): 43–64.

Mac Ginty, Roger. 2016. Political versus Sociological Time: The Fraught World of Timelines and Deadlines. In Arnim Langer and Graham K. Brown (eds), *Building Sustainable Peace: Timing and Sequencing of Post-Conflict Reconstruction and Peacebuilding*, 15–31. Oxford: Oxford University Press.

Misztal, Barbara A. 2016. Political Forgiveness' Transformative Potentials. *International Journal of Politics, Culture, and Society* 29(1): 1–18.

Nussio, Enzo, and Kimberly Howe. 2016. When Protection Collapses: Post-Demobilization Trajectories of Violence. *Terrorism and Political Violence* 28(5): 848–867.

Ramirez, Juliana. 2016. El No Ha Sido La Campaña Más Barata Y Más Efectiva de La Historia. *La Republica*, 5 October. www.larepublica.co/el-no-ha-sido-la-campa%C3%B1a-m%C3%A1s-barata-y-m%C3%A1s-efectiva-de-la-historia_427891.

Restrepo, Carlos. 2016. Lo Que Estamos Haciendo, No Es Para Hacerle Un Favor a La Guerrilla, Ni a Los Paramilitares, Ni Al Estado, Es Para Nosotros Mismos', Entrevista a Leyner Palacios | Organización Internacional Para Las Migraciones. *OIM Mision En Colombia*. 31 March. www.oim.org.co/news/%E2%80%9Clo-que-estamos-haciendo-no-es-para-hacerle-un-favor-la-guerrilla-ni-los-paramilitares-ni-al.

Ricoeur, Paul. 2004. *Memory, History, Forgetting*. Translated by Kathleen Blamey and David Pellauer. Chicago, IL: University of Chicago Press.

Rios Oyola, Sandra Milena. 2015. *Religion, Social Memory, and Conflict: The Massacre of Bojayá in Colombia*. Palgrave Studies in Compromise after Conflict. Basingstoke: Palgrave Macmillan.

Rivera, Norma. 2015. *Siempre Y Cuando*. Documentary. Bogotá, Colombia.

Rosa, Hartmut. 2005. The Speed of Global Flows and the Pace of Democratic Politics. *New Political Science* 27(4): 445–459.

Rosa, Hartmut. 2015. *Social Acceleration: A New Theory of Modernity*. Translated by Jonathan Trejo-Mathys. Reprint edition. New York: Columbia University Press.

Saffon, M. and R. Uprimny. 2009. Uses and Abuses of Transitional Justice in Colombia. In Morten Bergsmo and Pablo Kalmanovitz (eds), *Law in Peace Negotiations*. www.fichl.org/fileadmin/fichl/documents/Review_by_Abby_Zeith_in_Oxford_Socio-Legal_Review__Vol._1__2015___Issue_1.pdf

Sanchez, Gonzalo. 2015. Yo Estuve En … El Acto de Reconocimiento de Las Farc Ante Víctimas de Bojayá. Text. *ElEspectador.com*. 27 December. www.elespectador.com/noticias/nacional/yo-estuve-el-acto-de-reconocimiento-de-farc-victimas-de-articulo-607766.

Scheuerman, William E. 2004. *Liberal Democracy and the Social Acceleration of Time*. Baltimore, MD: The Johns Hopkins University Press.

Semana. 2016. Las Víctimas Votaron Por El Sí. 2 October. www.semana.com/nacion/articulo/plebiscito-por-la-paz-victimas-del-conflicto-votaron-por-el-si/496571.

Semana. 2017. Reforma Política de Fast Track a Slow Motion. *Reforma Política de Fast Track a Slow Motion*. www.semana.com/nacion/articulo/reforma-politica-a-paso-lento-en-camara-de-representantes/537385.

Walton, Oliver. 2016. Timing and Sequencing of Post-Conflict Reconstruction and Peacebuilding in Sri Lanka. In Arnim Langer and Graham K. Brown (eds), *Building Sustainable Peace*, 418–438. Oxford: Oxford University Press.

Zarama, Felipe. 2015. Caguan's and Havana's Peace Talks: Strategic Retreat or Stalemate Driven? In Cante Fredy and Hartmut Quehl (eds), *Handbook of Research on Transitional Justice and Peace Building in Turbulent Regions*, 471–508. Hershey, PA: IGI Global.

Zartman, I. William. 2000. Ripeness: The Hurting Stalemate and beyond. In Paul Stern and Daniel Druckman (eds), *International Conflict Resolution after the Cold War*, 225–250. Washington, DC: National Academy Press.

Zartman, I. William. 2001. The Timing of Peace Initiatives: Hurting Stalemates and Ripe Moments. *The Global Review of Ethnopolitics* 1(1): 8–18.

5 Anthropological reflections on violence and time in Argentina

Eva van Roekel

Introduction

In 2010 thousands of crimes against humanity that were committed during the military dictatorship (1976–1983) in Argentina were being prosecuted at several federal courts throughout the country. In July that year I was sitting with Laura Figueroa in a small office at the Ministry of Justice, Security and Human Rights in Buenos Aires.[1] Laura's father disappeared in the late 1970s and as a psychoanalyst she had been working with many survivors and relatives of the disappeared of the military dictatorship.[2] Laura was describing how many victims remember their violent pasts during the court hearings. She defined these processes of remembering not as chronological, yet as completely logical, in which decisive personal moments established important *antes y después* (before and after) in the lives of the victims. She also criticised the chronological organisation of historical evidence, based on a Western principle of linear classification and a progressive temporality. Victims' histories of the last dictatorship and its aftermath were rather a collective jumble of significant personal episodes that did not follow an established chronological thread of events. Although sometimes legally disqualified, for Laura these jumbles of shared memories were important evidence of the crimes.

I was talking with Laura as part of my research about the trials for crimes against humanity in Argentina, such as the thousands of illegal detentions and torture, the death flights, the disappearances and the illegal adoption of babies born in detention. Like many other conversations, our talk eventually descended into a conversation about the lived relationship between violence and time. Laura's explanations of the suffering were fused with Western psychological concepts of trauma and the therapeutic efficacy of catharsis (Hastrup 2003: 313). Yet the memory work Laura was talking about draws from a slightly different genealogy of self and suffering. Traumas and the continuous verbalisation of such were never reaching closure in the future. Traumas were considered intrinsic to being alive and remembering the past was based on a spiral understanding of time. Their traumatic pasts continuously intersected – although always in different forms – in the present.

During the same fieldwork in Argentina, I also talked with the perpetrators of state repression. These mostly military men had quite different conceptualisations of violence and time. Violence intrinsically belonged to human life and remembering the past belonged to a temporality that was much more focused on the future. Their memory work was concrete and stable and drew from a genealogy of self and suffering that was grounded in patriotism and the glory of combat. Although evaluations of past, present and future produced different temporalities for the victims and the perpetrators in Argentina, their experiences of violence and trauma both revolved around notions of perpetuation rather than transience. The violence of the 1970s and 1980s, for that matter, was not a social condition that was transitory in their lives.

Argentina's recent trials for crimes against humanity must be understood against a backdrop of forty years of social struggle for truth, justice and memory and against oblivion and impunity. These trajectories in post-conflict societies have often been coined as transitional justice. In brief, transitional justice addresses processes of accountability and reconciliation within political transitions from authoritarian regimes to more democratic societies. Transitional justice scholars have been mostly concerned with how the so-called healing of societies by means of truth, justice and memory practices can be understood as crucial phenomena to prevent violence breaking out anew, to establish a rule of law, and raise respect for human rights and democratic values in these countries. This debate mainly stems from two wide-ranging temporal approaches of how societies confront their violent histories – a backward-looking approach, which focuses mainly on retribution and reparations, and a forward-looking approach, which focuses primarily on restoration, healing and truth-telling (Barahona de Brito et al. 2001; Elster 2004; Kritz 1995; Roht-Arriaza and Mariezcurrena 2006; Teitel 2000). Many of these scholars acknowledge that in practice this division is less restrictive; in truth commissions and trials, for instance, these backward- and forward-looking approaches often intermingle.

Much of this literature is driven by a progressive and linear temporality, in which time progresses chronologically from one stage to another in series of subsequent events. 'Transition' even means passage or movement from one state to another (Oxford Dictionary 2015). Societies, then, would move from a state of violence to a state of non-violence. In everyday life the boundaries between violence and non-violence are often more slippery. Therefore, after more than two decades of theorising about these large-scale social processes in the aftermath of collective violence, to address the ongoing flows of formal and informal truth, memory and justice practices in different societies, the notion of 'post-transitional justice' has recently been introduced (Collins 2010; Davis 2013; Park 2014). Following this new perspective on transitional justice, understandings on accountability and reconciliation surpass limited analyses of political transitions from authoritarian regimes to more democratic societies. This shift towards post-transitional justice allows discussions on how underlying causes that affect transitional justice belong to long-term transformative processes of social, economic and political structures and relationships. To continue these productive discussions on the underlying social processes of transitional justice, I believe it is necessary to rethink the relationship between violence and time conceptually.

The anthropology of violence (Das 2007; Jackson 2002; Hinton 2005; Nordstrom and Robben 1995; Scheper-Hughes 1993) has increasingly contributed to the understanding of how people in conditions of post-conflict and structural violence experience their lives. These anthropologists have shown that 'violence is an inescapable dimension of people's existence, not something external to society and culture that 'happens' to people' (Nordstrom and Robben 1995: 2). Here I wish to contribute to this anthropological literature on violence and examine how local workings of time shape social processes of state violence, transition and reconstruction. I believe that this focus on time enables us to examine how transitional justice practices are not always linear movements away from violence and suffering, yet can also exist as ongoing moral practices to keep violence and suffering in the present.

Lived experiences of time

Assuming that the social world is historically conditioned, temporality is an important dimension (or structure) of how people experience the world in the here and now. Edmund Husserl, for instance, wrote that the now always has its horizon of past and future and is the source-point of all temporal positions (Brough and Blattner 2006: 127–128). In Argentina, however, I noticed different valuations of the past, present and future, which provided alternative contents to the lived experiences of time. In other words, temporality in everyday life is culturally dependent, dynamic and transformative. To understand how people actually experience past, present and future on their terms, I believe we should contrast the philosophical concept 'temporality' with the *lived* experience of time.

As suggested, people's lived experiences of time do not always fit a modern linear temporality. Time in the stories told by survivors of illegal detention and relatives of the disappeared suggests another temporal spin whereby certain lived events are part of a spiralled time frame. Time then takes place not in a series of identical cycles, yet it evolves spirally by rotating without repetition. Cultural constructions of temporalities can be arranged around varying relations between natural and social rhythms, cyclical and linear times, and repetition and irreversibility in order to understand how lived experiences of time shape people's everyday differently (Gell 1992). Individuals may even have different temporalities at different, or even intersecting, moments of their lives (Das 2007: 99). How people actually experience time is not one-dimensional.

To unravel lived experiences of time, phenomenologist Alfred Schutz has therefore suggested how 'lifeworldly time' is built up from 'subjective time', and intersects with the rhythm of the body as 'biological time', with the passing of the seasons as 'world time', and with organised calendars as 'social time'; we live in all these different temporal dimensions simultaneously (cited in

Schutz and Luckmann 1974: 47). To understand this multi-dimensionality of the living present, Maurice Merleau-Ponty in his reflections on temporality suggested that there exists an inseparability of time and the subject (Bakker 1969). *How* these different temporalities intersect is culturally and historically dependent. Clifford Geertz (1973: 389) argued that between a people's local conceptions of what it is to be a person and their local conception of the structure of history, there is an unbreakable internal link. I recommend, therefore, that lived experiences of time be less read as mere temporalities – as in how people relate to time in itself – than as cultural codifications of the self and the social within time. The analysis of how people define the self and the social in relation to local evaluations of the past, present and future facilitates deconstructing lived experiences of time. I think the below life histories of Emilio and Nora explain well how time operates in the everyday lives of many victims in Argentina and differs from a modern linear time that follows a progressive thread of clear events.

In 2010 Emilio Váldez introduced me to his grandmother Nora Ortega de Sánchez. Before our first get-together at home, Emilio had already told fragmentary and unrelated stories about how his uncle David Sánchez disappeared in 1977. At the same time, Emilio often expressed worries about his grandmother. She used to walk at Plaza de Mayo, the central square near the governmental palace in Buenos Aires, but since the late 1980s she had not been actively involved in the memory, truth and justice struggles in Argentina. Since his grandfather, Salvador Sánchez had died, Nora hardly got out of the house anymore. The first afternoon I met Nora we talked about her memories of David's disappearance. We also spoke about Emilio's political involvement in a local human rights organisation, and why he had moved in with his grandmother just a few months ago. Nora also worried a lot about Emilio, whose name she often confused with that of her son David. In bits and pieces Nora told me that after several fights with his mother and father-in-law, Emilio had been wandering the streets of Buenos Aires. Fear of also losing her grandson made her decide to let him into her home. While Nora continued her recollections about David, Emilio looked for papers and pictures that they had archived. The table was soon spread with papers that revealed only small parts of David's story. There were mostly Nora's personal letters to state officials asking for the whereabouts of her son. Filled with personal documentation and scarce official documentation, the kitchen table illustrated tangibly how a history of state violence turned into personal memories.

Like many victims in Argentina, Nora had created a small family shrine in her house: the pictures of David and Salvador stood in the middle, surrounded by fresh flowers and candles. Nora said that she preserved Salvador's ashes in an urn at a safe place. In an apologetic tone she continued, saying that David's pictures were the only things she had left. Jumping to the present, she still hoped that the forensic anthropologists would call and tell her that they found his remains. Although Emilio always urged Nora to contact the human rights lawyers that were pushing the trials' proceedings to see if they could prosecute for David's case too, Nora did not care much for a trial, she said; she just wanted him to lie at rest. When we sat at the kitchen table again, Nora recalled that they had agitated discussions with David at *this* table. Nothing important had been changed in the kitchen's furniture since that time. I noticed that it was somehow important that it was the same table. Nora further remembered their family conflicts in the kitchen in the seventies, and she reflected that they really had tried everything to stop David from getting involved in politics and the revolutionary struggle. She still felt wrong about this she once confessed while talking about her current indulgence towards Emilio, who had failed to complete his studies and was currently unemployed.[3]

The Ortega de Sánchez family never learned what had happened to their son. Survivors never saw him at a detention centre. David had vanished, like many disappeared in Argentina. The politeness of a last official return note represented the working methods of bureaucratic institutions during civil-military regimes and democracies. The return notes and meetings with officials were always given with bureaucratic polite words, but communicated practically nothing. Many families in Argentina depend on similar scarce information about the whereabouts of their loved ones. Because of this lack of knowledge, instead of one strong and coherent narrative, victims' recollections of the past depend on repetitive sharing of personal memories, pictures and notes. Nora continuously compared her own experiences and evaluations of the violence with those of other relatives with disappeared loved ones. This repetitive sharing of chaotic jumbles of personal memories and ongoing social negotiations about this past clearly does not produce a progressive temporality. The disappearances and other traumatic experiences from the past continuously spirals, although always in different forms, and vividly connects to victims' present lives. David's disappearance connects, time and again, with Nora and Emilio's present lives. Their social world does not consist of a coherent chronological narrative of knowable events, but rather of an ongoing spiral of ever-changing personal memories that connects the violence with the immediate present. The so-called transition from violence to non-violence works differently then.

I believe that the inability to conceive time in a non-linear and non-progressive temporality may be integral to an understanding of the self that is not only focused on human life that should be free from violence and suffering (Hastrup 2003), but is also based on a modern conceptualisation of experience that is orderly and knowable. Desjarlais (1996: 86–87) has also shown that social worlds do not always consist in the 'common' Aristotelian temporal order of experience that involves a cumulative layering of events and builds a greater whole than its parts. Instead of a strong and coherent narrative, time can also depend on momentary occupations and unconnected singular episodes. Argentine victims of state repression, who speak of their history as memory (Visacovsky 2007a), do not only produce messy temporal orderings in terms of decisive personal moments that establish important *antes y después* (before and after), but also produce life episodes that are inherently unknowable.

Irreparable ethical holes

I have argued that many people living in post-conflicts societies understand in a different manner how past, present and future are connected, a view that differs from the modern linear time frame quite common in transitional justice theories (see Hinton, Chapter 3, this volume). Studies on state repression in the Southern Cone of Latin America instead understand the violent past as ongoing and transformative social processes of memory making (Crenzel 2010; Jelin and Kaufman 2004; Klep 2012; Levin and Franco 2007; Van Drunen 2010). In Argentina, for instance, there emerged a view in which the past became an eclectic and conflictive repertoire of voices and was not based on the organisation of evidence according to the Western principle of linear classification and a progressive temporality (Visacovsky 2007a: 57–63). This repertoire of voices has produced a so-called 'subjective turn' due to the overall prevalence of the *testimonio* (testimony) as a legitimate approach to represent the violent past among Argentine victims of the last dictatorship (Sarlo 2005: 63). Victims' focus on the personal would be the method to construct authority and a way to legitimise the value of subjective memories (Oberti 2009: 73). Secrecy of the crimes committed, feelings of pain and guilt, and questions of individual and social responsibility regarding state violence have become enveloped in this mnemonic enactment of legitimacy. I suggest that this 'subjective turn' or 'particularization of the past' (Achugar 2008: 201–202) produces a particular unstable lived experience of time.

Although among victims there was not a clear and stable chronology of events, there was an unstable whole that was greater than its parts. Laura Figueroa, the psychoanalyst whose father disappeared, once reflected for instance that each intimate and lived truth of a victim contributed to the ongoing reconstruction of a social truth that exceeded the individual story.[4] There is a continuous oscillation between personal episodes and this larger, yet inherently contested, historical account that never reaches narrative closure. These social contestations, conflicts, and negotiations about what happened in the 1970s and 1980s are continuously being internalised, and produce a lived experience of time that is grounded in uncertainty. This uncertainty about the past goes further than the secrecy of how the military regime operated. Based on this intrinsic uncertainty, the notion that violence can be known in definite terms then becomes problematic.

As argued earlier, violence itself is also a slippery term (Hastrup 2003: 309). People give valuations to varying forms of violence, which also affects how and why people remember. Trauma or other shocking and violent life events may cause a rupture in people's temporality, as other studies have shown (Good 1994: 126–127). Caruth (1996) and Felman (2002) also understand this effect of trauma on temporality as a radical destruction or discontinuation between the past and the present, mainly because the traumatic event is seen as an event beyond experience and narration through which the past cannot be consciously incorporated in the here and now. However, in addition to their own interpretation of the violent event, the way people imagine the self produces how they define and experience time. When studying Punjabi women and their experiences of violence, Veena Das writes (2007: 99) about the co-existence of different temporalities regarding violence in the past that are intimately connected with everyday moralities – one that is frozen in the past and another that represents the flowing passage of life events. Based on local understandings of gender and suffering, certain memories of violence interweave in life as meaningful, while others cannot be easily revised through time (Das 2007: 88).

For Argentine victims of state repression, certain episodes of violence are valued as irreparable ethical holes in the self that need to be filled with references to moral values of justice and solidarity honouring the disappeared (Vezzetti 2010: 112–113). These irreparable holes, or traumatic marks, are exemplary and have converted remembrance into ongoing objects of obligation, through which the nature of the construction of the past completely changes (Vezzetti 2009: 23; Visacovsky 2007b: 285–286). Besides a political contextualisation of the secrecy of the crimes, and the lack of knowledge regarding those crimes, I suggest that these ongoing ethical performances concerning the irreparable holes shape how victims make sense of violence and time in their everyday. For these Argentines, the irreparable ethical holes are intrinsic to being alive, and continuously being reflective about them makes one human and distinct from other living organisms. The everyday is often understood as an unreflective moral disposition, which can be disrupted by moral breakdowns and reinstated if a subsequent liminal process of ethical performances is successful (Csordas 2013: 524; Zigon 2007: 140). For Argentine victims, however, it is exactly the opposite: returning to an unreflective moral disposition is unethical; instead, it is the act of ongoing reflexivity about irreparable ethical holes in their everyday that is considered to be moral (Van Roekel under review).

When I asked another Argentine psychoanalyst and survivor of illegal detention and torture how she treated these irreparable ethical holes, she explained that the human desire to put the unnamable holes into words has always existed; yet a closure or cure is intrinsically unrealisable.[5] According to her, this unachievable desire made human life meaningful. Those afflicted can thus live with trauma by continually working it through, yet keeping these ethical holes irreparable. I had heard many other Argentines talk about their traumas before, but never so clearly and cogently as at that time. Other Argentine survivors and relatives of the disappeared rather speak about *herida abierta* (open wound), which similarly renders violence in the past as a mode of present experience. After months in the field, I became more and more familiar with this mode of thinking about trauma.

A common medical understanding of an open wound is an injury that causes either an internal or external break in body tissue. In Argentina, this metaphor expresses local ideas about suffering based on violence in the past that resonate with the psychoanalytical idea of irreparable ethical holes.

Herida abierta not only refers to the last dictatorship: other violent events like the Malvinas War that killed more than 600 Argentine soldiers in 1982, the bombing of the Argentine Israelite Mutual Association (AMIA) that killed 85 people and injured hundreds more in 1994, and the death of almost 200 people in a fire at the nightclub República Cromañón in 2004, due to official negligence in monitoring safety measures, are considered *heridas abiertas* in local media and casual conversations.

This metaphorical understanding of an open wound is not restricted to Argentine victims. In the early twentieth century, Sigmund Freud in his article *Mourning and Melancholia* (1917) had already introduced a parallel between an open wound in body tissue and psychic suffering to address mourning processes. In the context of the last dictatorship, *herida abierta* particularly connects the violence of disappearances and illegal adoptions to notions of impunity and never-ending suffering in the present. Because of the lack of burials of the disappeared and the ever-existing possibility that a disappeared (grand-) child may be still alive, yet live with a different identity, survivors and relatives have not mourned properly. Therefore this kind of violence continuously intersects in the present as an open wound that never heals, or as an irreparable ethical hole that cannot be filled.

The next stories from military officers tell another Argentine story of the relationship between violence and time. Instead of an ever-changing and uncertain past that continuously spirals chaotically into the present, by institutionalising and materialising violence, time becomes an orderly and fixed business with emphasis on the future. These different lived experiences of time function intersubjectively and morally reinforce one and the other.

Chronicles of war

In September 2009, I met retired Lieutenant Colonel Leopoldo Cardoso who had fought in the 'war on subversion' himself. Leopoldo was confined to a wheelchair, 'a consequence of combat', he wryly said shortly after I arrived. Looking for printed evidence about the military actions he was to discuss, Leopoldo wheeled himself easily through the apartment. After reading some journals from the 1970s, I asked him how he was injured. Leopoldo narrated that in 1974 he had just started his military training, when the guerrilla movement PRT-ERP (Partido Revolucionario de los Trabajadores – Ejército Revolucionario del Pueblo) had entered the mountains of Tucuman. In early 1975, his commando unit was sent out to the edge of the Andean highlands in the northwest corner of Argentina for a counter-insurgency campaign. Leopoldo remembered that the area was full of PRT-ERP partisans. A few days after their arrival there was a violent confrontation between sixty military men and seventy *guerrilleros* (guerrilla members). In the same wry voice he said: 'And [there] I became how I am right now.' Leopoldo continued that a special helicopter unit transported him back to Buenos Aires and after nine months, in December 1975, he was rehabilitated. He then shouted: 'Nobody lives in the past!'

During the evening Leopoldo expressed similar general interpretations of the past: 'First there is the future, then there is present, and finally there is the past.' For Leopoldo people should always look forward first. Besides a scant personal account and his meticulous military analyses of the war against subversion, the entire evening Leopoldo expressed indignation about what he perceived to be the vindictive character of the current government, and victims' selective and ever-changing reinterpretations of the past. He reflected that the political violence of the guerrillas during the 1970s was even omitted from history on purpose. Leopoldo said bitterly to me:

> There may come a time that I have ended up in a wheelchair because I fell over a banana, as if I never have been in combat in Tucuman. I do not like that they change the history that I have lived.

Although Leopoldo acknowledged that he could not wipe out the violent past, like other military officers did, he repeatedly condemned the current re-writing of history by victims and state officials, regarding the trials for crimes against humanity and the new official educational programmes and commemorative practices about what had happened during the last dictatorship.

After this talk with Leopoldo, I heard many other military officers interpret the past like this. Their lived experiences of time seemed more chronological and organised by combat events. At the courts, officers' allegations were often about troop movements and military analyses about the emergence of the different guerrilla groups in Argentina and Latin America. Although never entirely omitted, the personal and the affective seemed less relevant in their stories. These officers related to violence and time in very different ways than the victims I had been talking to. There was no spiral of traumas and continuous re-interpretation of bits and pieces of an uncertain past. They even condemned the continuous 're-writing of history'. Above all, there seemed little psychological depth in their stories. Ex-General Jorge Rafeal Videla's historical account, in this sense, was exemplary.

We spoke several times during his pre-trial detention at a military prison outside Buenos Aires. Asking about what had happened and why he was in prison, Videla gave a brief institutional analysis about the legitimate transfer of power from a democratically elected government to the armed forces. According to Videla, 'they knew it would be war'. Then Videla mentioned the war on subversion. He reflected that they had won the war militarily, but had lost it politically. Finally, Videla remembered the junta trial in 1985 when the junta stood trial for kidnapping, torture and murder, but not for a systematic plan to snatch babies. With disapproval, Videla added that only after several years the human rights groups transformed these adoptions into a systematic plan. I still remember that Videla's narrative was confident, determined and clear, and there was little opportunity to discuss alternative interpretations.

A study of military memory processes in the Uruguayan Army has shown similar determinate discourses about the violence and argues that, through resources of objectivity, these officers create a tolerable self-presentation (Achugar 2008: 122–123). Achugar defines these resources of objectivity as clear historical references and historical comparisons, with a focus on events and not on participants. She suggests that Uruguayan military officers have a strong tradition of producing such institutional factual accounts that document events and circumstances, attempting to give causal explanations and fixed chronologies of the past (Achugar 2008: 200–201). Such institutional and detached historical narratives are said to be also the preferred genre among military officers who participated in the state repression in Brazil (Huggins 2000: 68–69). This 'professionalism discourse' particularly disembodies violence by substituting nonhuman organisational factors for human agency and physicality and by justifying violence through dispassionate calculus. Huggins uses this detached professionalism approach to explain a strategic military method to circumvent accountability. However, I suggest that a certain insistence on professionalism in military memories is a sign not only of avoidance of accountability and positive self-representation but also of an institutional enshrinement of order and loyalty. In other words, the cultural codifications of the self and the social and evaluations of the past violence from a military perspective also define how these men experience time as lived. Perhaps precisely because of a military professionalism discourse, the stories of the perpetrators I spoke with were often very physical and passionate military narratives of violence.

The materiality of violence

Military officers I spoke with almost never framed their experiences of violence and death in psychological terms that belong to a modern suffering self. The stories of combat that many officers told were rather concrete bloody events with few references on their subjective or emotional states. Colonel Alfonso Palacios' chronicles of war, I think, are a good illustration of how institutional narratives of violence do not have to be detached and disembodied; they are only more focused on the physical than the psychological, and the social than the individual aspects of experience.

Retired Colonel Alfonso Palacios had been an active military officer between 1976 and 1983. Until the time of writing, he has not been held responsible for crimes against humanity, like many of his fellow officers I met in military prison in 2010. When I asked Alfonso to reflect on his personal experiences during the 1970s until the early 1980s, he promptly began with the year 1972, which was the year when he graduated from the military academy.[6] Alfonso recounted that afterwards he was assigned to an artillery group located in the province of Buenos Aires. Then in January 1974 he got his baptism of fire, when the guerrilla movement PRT-ERP attacked their military barracks. That night, according to Alfonso, more than 120 men entered the barracks, while 120 others blocked the entrance. They returned fire and apprehended several *guerrilleros* (guerilla members). After fighting the entire night, the battle ended with three casualties.

After this personally detached historical overview Alfonso asked rhetorically: 'How did I experience that night?' He remembered that they heard a gunshot, which, at first, they thought was an accident. Then they heard two gunshots, followed by machinegun fire and several explosions. Alfonso narrated that then they knew that something serious was happening. When I asked Alfonso again how these events had marked his own experiences of the past, Alfonso replied that they had already been living with the war against terrorism for some time. At that moment, the armed forces had received political instructions not to intervene. After another detailed account of how a federal chamber to combat terrorism with national jurisdiction came into being during the military government of General Lanusse in 1971, somehow sensing I was searching (mistakenly I guess) for a different answer, Alfonso repeated my question: 'How did it mark me?'

Alfonso then vividly remembered how he had lifted the bodies of a dead soldier and a dead commander-in-chief, and how he had held the dying wife of the commander-in-chief in his arms. Apologising for his graphic descriptions, Alfonso narrated how his own large hand entered the exit wound of the woman's back. His hand was full with bloody pieces of human flesh. He reflected that his feelings were strangely mixed – satisfied that they had managed successfully to overpower the guerrilla attack, yet pain for the human lives that had been lost. Then he repositioned his reflection and said determinedly: '... to be completely honest, the main sensation was that we had accomplished our mission'. At other times, I asked Alfonso about the violence they committed. Like his former commander-in-chief Videla, his accounts were always confident institutional histories and internally consistent, yet one afternoon when I asked why they *deliberately* disappeared their opponents he stumbled, 'Well ... what do I know ... I do not know ... I should think this over ...' His silence lasted for a few seconds and then Alfonso quickly resumed with his institutional analysis of the revolutionary violence and the political and economic interest of the figure of the disappeared in Argentina. Although Alfonso was a product of long military socialisation and military morals, he had suddenly lost his confidence and professional detachment.

Das (2007: 95–97) writes that, after violence, we should look at how victims are immersed in time as lived that shape the contours of their lives. I suggest that this also counts for those that perpetrate the violence. Most of time, I noticed that many military memories of violence interweave in everyday life as meaningful, while others could not be easily revised through time. I want to look here particularly at those military memories of violence that interweave in life as meaningful. Alfonso's memories of the violence presumes that time enfolds in chronological ways with a course of events that are characterised by a certain stability and materiality. Like many other conversations, our talk had descended into a conversation about the relationship between violence and time. Officers' explanations of the violence were almost completely void of any dominant Western psychological concepts of trauma, even when talking about the violence that was difficult to revise through time. Their memory work drew from a different genealogy of self and suffering.

Again, such lived experiences of time should be read less as mere temporalities than as alternative cultural codifications of the self and the social in relation to valuations of time. In this social world time is infused with order and loyalty, and violence is not a transient condition in military everyday life. Military officers explained to me that executing violence is a vocation and a lifelong duty. For these officers, violence is thus not only a normal condition of the military self, but also an ongoing moral practice through which they differentiate themselves from civilian life. This does not mean that officers consider all forms of violence moral; some acts of violence were inexplicable, even after 40 years. Among many military officers, however, there was no desire to put the inexplicable into words, as it did not make their lives more meaningful. Instead many believed that continuously delving into the past was harmful (Van Roekel 2016: 146–152).

Without claiming that one mode is the accurate or only mode in which lived experiences of time are produced, I believe the stories told here show how different and dynamic forms of time as-lived and valuations of violence co-exist and produce different understandings of local transitional justice trajectories. The counter memory practices of military men that I discuss next also belong to these long-term transformative social processes in the aftermath of collective violence.

Remembering loyalty

If violence transforms how people experience the progression of time, it does so, I suggest, less as a radical destruction or discontinuation of people's temporality, than as the institutionalisation, or socialisation, of a particular imaginary of the past. Achugar (2008: 206), for instance, has argued that some military officers in Uruguay have acknowledged the possibility of having a plurality of truths and interpretations of the past, which has been an important change in their institutional rhetoric. Temporalities, or how people relate to time, are always the product of social change. Other studies on military memory processes about the repression in the Southern Cone have shown that the military narrative of having saved the nation from internal aggression may have evolved over time, yet based on the principles of a national security doctrine, essential messages have not altered (Hershberg and Agüero 2005: 32–33). For men like Alfonso and Leopoldo, the essential messages of their violent histories also did not vary much over time (Robben 2005). Yet their stories of bloody combat do contest a dominant essentialist approach to military recollections of history.

Gradually I had come to know that many military officers, their wives and children reason that because of the continuous human rights' struggle for memory, truth and justice for the disappeared, and particularly since the 'human rights politics' of the Kirchner governments (2003–2015), there had emerged a too selective remembrance of the violence perpetrated by the state. Different social groups loyal to the armed forces, like FAMUS (*Familiares y Amigos de los Muertos por la Subversión*), have instead publicly remembered acts of political violence perpetrated by the guerrillas during the 1970s (Salvi 2012). Several local organisations, like CELTyV (*Centro de Estudios Legales sobre el Terrorismo y sus Víctimas*), continue to work on the transmission of these 'complete memories' or 'complete histories' and struggle for the recognition of victims of terrorism.[7] This counter memory must be understood against a backdrop of decades of anti-military sentiment in Argentina. Since 1983, the previously influential Argentine Armed Forces have been progressively pushed to the margins of the state (Frederic 2010: 229–230; Van Roekel 2016: 136–138). By retelling stories of guerrilla violence, like the kidnapping and subsequent death of Colonel Argentino Del Valle Larrabure, or the bombing of the apartment of Vice-Admiral Armando Lambruschini, in which his 15-year-old daughter Paula died, Valentina Salvi (2012: 270–275) argues that the military and these counter memory organisations follow a personalised approach to the violence which similes victims' memories of the last dictatorship.

Those who understand military counter memory work strictly in accordance with a modern understanding of the suffering self have theorised that officers' civilian roles as fathers and husbands are particularly emphasised, which would fit better in the cultural image of a victim than the image of a courageous military officer who died in combat (Salvi 2012). These images of the suffering military make sense as strategic public commemorations. The values and interests in military commemoration and ritual and its interpretations of it reveal also something different. By retelling these stories of combat and fallen soldiers, what the military officers were really getting was something else – respect. Here military culture differs from a modern approach to the representation of violence wherein the public rendering of the fallen soldier needs a more institutional reading of military memory work. If not only strategic political processes of victimisation, retelling stories of fallen soldiers, I suggest, are also military enactments of loyalty and deference that turn remembrance into ethical objects of military obligation. Ex-Major Aldo Domínguez' account on the interaction between military loyalty and time is illustrative.[8]

Aldo often spoke about his participation in the *Carapintadas* (Painted Faces).[9] They rebelled against the legal accusations of murder and torture against fellow officers in the late 1980s and the destruction of the military capacity to defend the homeland due to financial cuts. He lost his commission during the last military uprising 'Operation Dignity'. Aldo reflected on the reason he had rebelled against his superiors, and which had managed to stop the prosecution of his fellow officers. Even though they had been in military prison for the rebellion, Aldo still reasoned that they had done the right thing. He was proud of what they had done and spoke with great deference for those who had lost their lives for the good cause. Lacking official recognition for those fallen soldiers, they had created their own commemorative rituals. Nevertheless, for Aldo, these last military uprisings had marked the definite endpoint in Argentine military history and reflected that,

> [After our last rebellion], the Armed Forces soon stopped to be a real threat to the executive power. From then on, the enduring destruction of the Armed Forces really consolidated.

Aldo's memories of the rebellion continuously oscillated between their actions, their military hierarchy and their protective duty to the nation. For Aldo these three realms, self, institution and the nation, were not separate entities, yet rather 'one meaningful totality' (Merleau-Ponty in Bakker 1969: 72) that were tied together by rituals that manifested military loyalty and service to their fellow men, the armed forces, and the fatherland. Aldo's memories celebrated the greatness of this fatherland.

Argentine cadets are still educated to celebrate the glory of the nation and military loyalty. Although new military attitudes among the young cadets exist, which are more democratic, flexible and individual, the protagonism and persistence of a 'traditional cosmology' continues to shape a shared military identity (Badaró 2009: 217). Institutional rituals like daily chanting of the national anthem and raising the national flag produce ideas of military transcendence and mythical times of the fatherland where the glorified past should always be revivified in the present, and projected into the future (Badaró 2009: 185–222). These military rituals partially generate the internalisation of loyalty. Repeatedly telling stories, about Colonel Argentino Del Valle Larrabure, answers this need for ongoing re-enactments for those that have fallen in combat. These ritual commemorations are not only manifestations of military victimisation in the past, but also constitute expressions of patriotism and respect that are projected into the future.

Nonetheless, there is not a single monolithic military narrative of the dictatorial past: military officers' lived experiences of time, with their professional discourses of war, a focus on the future, and multiple stories of bloody combat and fallen soldiers mostly depend on stability, chronology, solidity, homogeneity and coherence. So-called resources of objectivity, like historical references, cause and effect relations and events, dynamically interact with military notions of self that are tied to the institution and the nation. Although this produces a more coherent and organised stance to time as lived, with a focus on the future, it does not mean that these military men move away from the violence. Violence is intrinsically part of military life, and those fallen should be remembered as ongoing acts of military obligation.

Conclusions: Transitional justice and temporality

What I have argued here is that conceiving of time as linear and progressive, and violence and suffering as social conditions that are transitory, is unproductive to further our understanding of how and why societies deal with truth and justice in the aftermath of collective violence in the various ways they do. I have suggested that it is in the specific nature of how communities and individuals conceive the relationship between violence and time that we can start to rethink transitional justice processes. In other words, not only do lived experiences of time depend on different conceptualisations of the self and differing valuations of the past, present and future, they also produce multiple forms of memorialisation and multiple meanings and outcomes of retribution and reparation in transitional justice settings. To continue thinking in the common temporal categorisation of backward-looking and forward-looking approaches in transitional justice literature would therefore do injustice to the complex workings of the multiple co-existing lived experiences of time of many people living in post-conflict settings. These findings resonate with the discrepancies between dominant and subaltern temporalities addressed by Richard Benda (Chapter 8) in the case of Rwanda.

On the one hand, in the case of Argentine victims of state repression, instead of one modern linear time frame, in which time progresses from one stage to another in series of unambiguous events, their lived experiences of time rather develop in spiralled notions of an uncertain and personalised past into the unstable present. In other words, how people conceive past does not have to be knowable and orderly. Victims' traumatic pasts continuously intersects – albeit differently – in the present. Time then takes place not in a series of identical cycles, yet it evolves spirally by rotating without repetition. Based on a psychoanalytical understanding of self, practices of remembrance and retribution of the inflicted traumas become ongoing ethical performances, which change the nature and relationship with the violent past, and violence and time in general. Traumatic suffering, then, is not transient; it is intrinsic to human life.

On the other hand, for the perpetrators of the Argentine conflict time enfolds in more chronological ways with a course of factual evens that is characterised by a certain stability and materiality. Time becomes an orderly and fixed business with a focus on the future. Their commemorative rituals are not only manifestations of military victimisation in the past, but also institutional glorifications of future nation. However, these lived experiences are not detached and disembodied; they are only more focused on the physical than the psychological, and the social than the individual aspects of experience. In this social world time is infused with order and loyalty, where violence is likewise not a transient condition in their everyday life; it fundamentally belongs to military life.

Thus, instead of a single linear time frame, Argentine lived experiences of time oscillate between chronology and chaos, the personal and the political, and the personal and the institutional. Instead of practices that would leave violence in the past (backwards or forwards), I have suggested that the truth, memory and justice practices of the past forty years in Argentina are rather ongoing re-enactments of the violence and the suffering that intersect with varying local definitions of self and the social. Cultural understandings of why and how societies and persons recover differently from collective violence should therefore always begin with this analytical step back, and start with the question how people conceive of what is to be a person connects with their local conception of the structure of history (Geertz 1973: 389). This is what I have called the *lived* experience of time.

If we acknowledge that this cultural approach to violence and time is more constructive, this new conceptualisation enables us to focus our attention on how communities and individuals actually conceive transitional justice. This is where we can see how the 'universal' temporal order of transitional justice is not a precondition for how people live with previous atrocities and human rights abuses. By contrast, a cultural approach to violence and time, potentially lends itself more easily to further our understanding about large-scale social processes concerning atrocities and human rights abuses that are often coined as transitional justice.

Notes

1 Interview with Laura Figueroa, 2 July 2010, Buenos Aires.
2 All the names in this chapter are pseudonyms to protect informants' privacy.
3 Interview with Nora Ortega de Sánchez, 8 October 2010, Buenos Aires.
4 Interview with Laura Figueroa, 2 July 2010, Buenos Aires.
5 Interview with Carla Ávila, 21 July 2010, Buenos Aires.
6 Interview with Alfonso Palacios, 22 October 2010, Buenos Aires.
7 FAMUS is no longer active. See for information on CELTyV and complete memory on website. Retrieved 15 November 2015 www.celtyv.org/.
8 Interview with Aldo Domínguez, 22 June 2010, Buenos Aires.
9 The *Carapintadas* were a group of dissident military officers of the Argentine army who took part in the uprisings during the presidency of Raúl Alfonsín in Argentina in the late 1980s.

References

Achugar, Mariana. 2008. *What We Remember: The Construction of Memory in Military Discourse*. Amsterdam: John Benjamins Publishing.

Badaró, Máximo. 2009. *Militares o ciudadanos. La formación de los oficiales del Ejército Argentino*. Buenos Aires: Prometeo Libros.

Bakker, R. 1969. *Kerngedachten van Merleau-Ponty*. Roermond: Romen & Zonen.

Barahonade Brito, Alexandra, Carmen González Enríquez and Paloma Aguilar (eds). 2001. *The Politics of Memory and Democratization: Transitional Justice in Democratizing Societies*. Oxford: Oxford University Press.

Brough, John B. and William Blattner. 2006. 'Temporality'. In *A Companion to Phenomenology and Existentialism*. Hubert L. Dreyfus and Mark A. Wrathall (eds). Malden, MA: Blackwell Publishing, 127–134.

Caruth, Cathy. 1996. *Unclaimed Experience: Trauma, Narrative, and History*. Baltimore, MD: Johns Hopkins University Press.

Collins, Cath. 2010. *Post-transitional Justice: Human Rights Trials in Chile and El Salvador*. University Park: Pennsylvania State University Press.

Crenzel, Emilio. 2010. *Los desaparecidos en la Argentina: memorias, representaciones e ideas 1983–2008*. Buenos Aires: Biblos.

Csordas, Thomas J. 2013. 'Morality as a cultural system?' *Current Anthropology* 54(5): 523–546.

Das, Veena. 2007. *Life and Words: Violence and the Descent into the Ordinary.* Berkeley: University of California Press.

Davis, Coreen. 2013. *State Terrorism and Post-Transitional Justice in Argentina: An Analysis of Mega Cause I Trial.* New York: Palgrave Macmillan.

Desjarlais, Robert R. 1996. 'Struggling Along'. In *Things as They Are: New Directions in Phenomenological Anthropology.* Michael Jackson (ed.). Bloomington: Indiana University Press, 70–93.

Elster, Jon. 2004. *Closing the Books: Transitional Justice in Historical Perspective.* Cambridge: Cambridge University Press.

Felman, Shoshana. 2002. *The Juridical Unconscious: Trials and Traumas in the Twentieth Century.* Cambridge, MA: Harvard University Press.

Frederic, Sabina. 2010. 'Fotografías de la configuración profesional de los militares en el contexto de su declinación come elite estatal'. In *Las prácticas del Estado. Política, sociedad y elites estatales en la Argentina del siglo XX*. Mariano Ben Plotkin and Eduardo Zimmermann (eds). Buenos Aires: Edhasa, 210–233.

Freud, Sigmund. 1917. 'Mourning and Melancholia'. In *The Standard Edition of the Complete Psychological Works of Sigmund Freud, Volume XIV (1914–1916): On the History of the Psycho-Analytic Movement, Papers on Metapsychology and Other Works.* London: Hogarth Press, 237–258.

Geertz, Clifford. 1973. *The Interpretation of Cultures: Selected Essays.* New York: Basic Books.

Gell, Alfred. 1992. *The Anthropology of Time: Cultural Constructions of Temporal Maps and Images.* Oxford: Berg.

Good, Byron J. 1994. *Medicine, Rationality, and Experience: An Anthropological Perspective.* Cambridge: Cambridge University Press.

Hastrup, Kirsten. 2003. 'Violence, suffering and human rights: Anthropological reflections'. *Anthropological Theory* 3(3): 309–323.

Hershberg, Eric and Felipe Aguëro. 2005. *Memorias militares sobre la represión en el Cono Sur: Visiones en disputa en dictadura y democracia*. Madrid: Siglo XXI.

Hinton, Alexander L. 2005. *Why Did They Kill? Cambodia in the Shadow of Genocide.* Berkeley: University of California Press.

Huggins, Martha K. 2000. 'Legacies of authoritarianism: Brazilian torturers' and murderers' reformulation of memory'. *Latin American Perspectives* 27(2): 57–78.

Jackson, Michael. 2002. *The Politics of Storytelling: Violence, Transgression and Intersubjectivity.* Copenhagen: Museum of Tusculanum Press.

Jelin, Elizabeth and Susana G. Kaufman. 2004. 'Layers of Memory: Twenty Years after in Argentina'. In *Commemorating War: The Politics of Memory.* Timothy G. Ashplant, Graham Dawson and Michael Roper (eds). New Brunswick, NJ: Transaction Publishers, 89–110.

Klep, Katrien. 2012. *Transitional Justice and Commemorative Practices: Processes of Memorialisation in Chile.* PhD dissertation, Department of Cultural Anthropology, Utrecht University.

Kritz, Neil J. (ed.) 1995. *Transitional Justice: How Emerging Democracies Reckon with Former Regimes.*Volume I: *General Considerations.* Washington, DC: USIP Press.

Levin, Florencia and Marina Franco. 2007. *Historia reciente: perspectivas y desafíos para un campo en construcción.* Buenos Aires: Paidós.

Nora, Pierre. 1989. 'Between memory and history: Les Lieux de Mémoire. Special Issue: Memory and counter-memory'. *Representations* 26: 7–24.

Nordstrom, Carolyn and Antonius C.G.M. Robben. 1995. *Fieldwork under Fire: Contemporary Studies of Violence and Survival.* Berkeley: University of California Press.

Oberti, Alejandra. 2009. 'Memorias y testigos, una discusión actual'. In *Memoria(s) y política. Experiencia, poéticas y construcciones de nación.* Maria del Carmen de la Peza, ed. Buenos Aires: Prometeo Libros, 67–86.

Oxford English Dictionary. 2015. Oxford: Oxford University Press.

Park, Rebekah. 2014. *The Reappeared: Argentine Former Political Prisoners.* New Brunswick, NJ: Rutgers University Press.

Robben, Antonius C.G.M. 2005. *Political Violence and Trauma in Argentina.* Philadelphia: University of Pennsylvania Press.

Roht-Arriaza, Naomi and Javier Mariezcurrena (eds). 2006. *Transitional Justice in the Twenty-first Century: Beyond Truth versus Justice.* Cambridge: Cambridge University Press.

Salvi, Valentina. 2012. 'Sobre "memorias parciales" y "Memoria Completa". Prácticas conmemorativas y narrativas cívico-militares sobre el pasado reciente en Argentina'. In *Topografías conflictivas. Memorias, espacios y ciudades en disputa.* Anne Huffschmid and Valeria Durán (eds). Buenos Aires: Nueva Trilce, 265–280.

Sarlo, Beatriz. 2005. *Tiempo pasado. Cultura de la memoria y giro subjetivo. Una discusión.* Buenos Aires: Siglo Veintiuno.

Scheper-Hughes, Nancy. 1993. *Death without Weeping: The Violence of Everyday in Brazil.* Berkeley: University of California Press.

Schutz, Alfred and Thomas Luckmann. 1974. *The Structures of the Life-World.* Richard M. Zaner and H. Tristan Engelhardt Jr., trans. London: Heinemann Educational Books.

Teitel, Ruti G. 2000. *Transitional Justice.* Oxford: Oxford University Press.

Van Drunen, Saskia. 2010. *Struggling with the Past: The Human Rights Movement and Memory in Post-Dictatorship Argentina (1983–2006).* PhD dissertation, CEDLA Department of Latin American Studies, University of Amsterdam.

Van Roekel, Eva. 2016. *Phenomenal Justice: State Violence, Emotion, and the Law in Argentina.* PhD dissertation, Department of Cultural Anthropology, Utrecht University.

Van Roekel, Eva. Under review. 'Traumatic home: Argentine victimhood and the everyday moral comfort of trauma'. *Ethos, Journal of the Society of Psychological Anthropology.*

Vezzetti, Hugo. 2010. 'Memoriales del terrorismo del estado en Buenos Aires: representacion y politica'. In *Memorias urbanas en dialogo: Berlin y Buenos Aires.* Peter Birle, Vera Carnovale, Elke Gryglewski and Estela Schindel (eds). Buenos Aires: Buenos Libros, 101–117.

Vezzetti, Hugo. 2009 *Sobre la violencia revolucionaria: memorias y olvidos.* Buenos Aires: Siglo Veintiuno.

Visacovsky, Sergio E. 2007a. 'Cuando las sociedades conciben el pasado como 'memoria': un análisis sobre verdad histórica, justicia y prácticas sociales de narración a partir de un caso argentino'. *Antipoda* (enero-junio): 49–74.

Visacovsky, Sergio E. 2007b. 'Historias próximas, historias lejanas. Usos sociales de las distancias temporales en la organización de las experiencias sobre el pasado: El caso del servicio de psiquiatría del Lanús'. In *Historia reciente. Perspectivas y desafíos para un campo en construcción*. Marina Franco y Florencia Levín (eds). Buenos Aires: Paidós. 279–305.

Zigon, Jarrett. 2007. 'Moral breakdown and the ethical demand: A theoretical framework for an anthropology of moralities'. *Anthropological Theory* 7(2): 131–150.

6 Negotiating temporalities of accountability in communities in conflict in Africa

Victor Igreja

Introduction

This article focuses on how war survivors and their offspring negotiate temporalities of accountability in communities in conflict in Mozambique. Specifically, the analysis dwells on serious violations that were committed between 1971 and 1974, during the Portuguese colonisation, and civil war (1976–1992) between the Frelimo government and the Renamo rebels in Gorongosa, a district in the centre of the country. The notion of communities in conflict refers to the existence of various participants and dynamics of unrest in the same national political space. On the one hand, the main political rivals in Mozambique, the Frelimo and Renamo parties, were involved in serious political skirmishes for nearly two decades (1994–2012) following the official end of the civil war. These conflicts were fuelled by political intolerance and lack of reconciliation, and manifested through mutual use of memories as weapons to accuse one another of perpetrating serious crimes during the civil war (Igreja 2008). These unresolved conflicts were never subjected to any formal accountability, and instead triggered new military confrontations which heightened conflicts and insecurity in the centre of the country. On the other hand, young people experience severe difficulties and are sometimes denied agency to participate in processes of social reproduction. Following the civil war, many survivors wished to forget the horrors and move on. They were instead confronted with traces of past violations which surfaced in the household sites through acts of *ku gurukuta* (unburden murmur), mutual accusations and gossips, which nevertheless remained discreet and veiled to non-kin members.

The traces of the past violations also surfaced in more noticeable ways through the disruptive intrusions of *gamba*, which are spirits of soldiers who died during the civil war. In the first decade after the end of the civil war (1992–2002), struggles for accountability were dominated by the intrusions of *gamba* spirits who had high degree of specificity to serious offences perpetrated during the civil war. Over time the configuration of the *gamba* spirits changed to adopt multiple temporalities and denoting any individual presumed to have died from an unjust death in the past (Igreja 2012a). Thus, the notion of temporalities of accountability refers to the acceptance or rejection to pursue liability for serious violations committed over an indeterminate time span. In cases of rejection, for example, the imposition of a time limitation bars possibilities of liability based on the belief that the material evidence of violations becomes meaningless outside the predefined timeframe of reference.

In turn, in cases of acceptance, as in the Gorongosa region, and many other locations in Africa (cf. Douglas 1980), liability over time is sustained by a belief that 'micero ai vundi', which means that unresolved serious conflicts do not rot or fade away with passage of time given the involvement of spirits of the dead in such conflicts (Igreja and Skaar 2013). Ethnographic records from numerous communities in Africa and around the world show how spirits are regarded as disembodied figures but also as persons with volition of their own and capable of influencing the lives of the living (Igreja 2009; Lambek 1993; Masquelier 2001). Often, spirits are inseparable from specific historical circumstances of death (Kwon 2008) and these disembodied agents actualise themselves in the present and future through the bodies of the living, which are conceived as sites acquiescent for enduring afflictions, protracted social conflicts and possession.

In this regard, when the *gamba* spirits suddenly take possession of a person's body in Gorongosa, this signals that time for *ku gurukuta* is up and accountability for past serious violations has been unleashed in the open. *Gamba* spirits often possess the bodies of young people which seriously hamper the hosts' capacities to bear children and participate in other meaningful social activities. Yet the same intrusive *gamba* spirits is also an agent of authority that drives the possessed person to indict those believed to be responsible for serious violations in the past. It is the failure to come to terms with these violations that triggers and sustains the afflictions that torment their lives (Igreja 2014a). Thus, embodied accountability revolves around the bodies of survivors and their offspring both as sites of torment and sites of accountability, in tandem with polydiscursivity encompassing songs of war, loss and suffering, pithy locutions of ritual testimonies, indictments, confrontations and defences, narratives and metaphors. All these pave the way for transforming abstract accusations into reality, locally termed as *ku dembula*, which means to acknowledge the person's wrongdoings. Once liability is established, they enact *ku lapiwa* (to pay back), which is specifically used for paying back in cases of violent deaths.

It is meaningful to set out the contrasts and overlaps of embodied accountability with official processes of accountability encapsulated in the notion of transitional justice. It overlaps in that it confirms the existence of multiple sites of accountability and diverse actors and mechanisms involved in processes to come to terms with legacies of serious past wrongs. The contrast, in turn, highlights the flexibility of embodied accountability which draws on diverse serious wrongdoing, irrespective of the passage of time, number of alleged victims, social position of alleged perpetrators as well as regardless of the identification of the bodies of the dead. The flexibility of embodied accountability is exemplary beyond the everyday life of the Gorongosas as it provides further impetus to the notion of 'era of the witness' (Wieviorka 2006) by expanding the temporal scope and temporal rhythms (Mueller-Hirth 2017) to opening up and account for serious injustices that have been forgotten, distorted or hidden by diverse social actors over time.

These ideas will be fleshed out through a case study of a conflict involving a family of survivors of the civil war (1976–1992) in Mozambique. The process of resolution of a serious violation perpetrated in the civil war unexpectedly also required dealing with a serious violation that had unfolded sometime during the Portuguese colonisation and had implicated members of the same family group. The acceptance to give voice to and hear the alleged violations, despite their long-standing existence, paved the way for the resolution of the conflict.

Temporalities of global and local accountability

In the twentieth and twenty-first centuries, reports of colonial exploitation, Nazism, genocide, apartheid and dictatorships of the right and left and the violence that unfolded in numerous post-colonial civil wars were technically labelled crimes against humanity, war crimes or serious violations. This nomenclature is part of an evolving mainstream platform of 'global legalism', which involves excessive faith in the efficacy of international law to resolve conflicts (Posner 2009: xii). Earlier in the 1990s, transitional justice emerged as a specialised field that analyses how official mechanisms are deployed in multiple sites of accountability – such as criminal trials, administrative justice measures, truth commissions, amnesty laws, community rituals – to deal with legacies of mass political violence in a specified timespan (Hayner 2001). Yet tensions prevail over how to attain accountability in ways that are also meaningful for the diversity of people victimised by an array of serious violations over time.

Official approaches to accountability – both in their retributive and restorative versions – are based on mechanistic mindsets, specific notions of personhood, and a number of bureaucratic and quantifiable techniques. The Rome Statute focuses on 'widespread or systematic attack' directed against civilian populations in a given space and time which might become of interest for the legal specialists of global accountability (ICC 1998: 4–10). Time also shapes the modalities of accepted jurisdiction for legal prosecution: alleged serious crimes perpetrated before the ICC establishment are regarded, at least legally, as invalid therefore excluded from accountability, whereas those perpetrated after the court's formation can be potentially included within its jurisdiction. Additionally, alleged crimes committed before or after the age of criminal responsibility as legally defined (18 years old) receive differential treatment (ICC 1998: 18). The enforcement of this latter notion is to the detriment of other cultural conceptions of age, which, unlike chronological time, focus on social actions as determinant of the social position of the individual (Igreja 2015b; Last 2000).

Despite the various criticisms of clocktime perspective (Leccardi 1996; Munn 1992), the conception of time as 'before and after' remains influential and shapes broader ideas about intentionality, memory, narrative, history and justice. The ensuing justice centres on those regarded as most responsible for, or masterminds of, high rates of violent deaths. Although the focus on individual accountability has also been questioned, legal responsibility requires that 'either bodies must be uncovered from mass graves and identified' or 'the number of deaths must be otherwise convincingly established' (Hagan, Schoenfeld and Palloni 2006: 334). The habit of identifying reality with presence is consistent with a certain theory of time prevalent in the West whereby time is conceived as the chronological succession of instants (Al-Saji 2004: 204). Events in this type of reasoning are also considered to occur in linear ways, one after another, which creates the illusion of assertively grasping events and their contours strictly defined and measured in the flow of time (Munn 1992). It is expected that this approach can seal off the past and initiate a new era, although the attainment of this goal is a matter of debate.

This approach to time and ensuing legal practices of accountability is not, however, universally shared. In various non-Western communities, notions of temporality lean toward the idea of 'non-chronological time' (Deleuze 1989: 82). In such cases 'the past and the present do not denote two successive moments, but two elements which coexist' (Deleuze 1991: 59) and are constituted as diverse activities unfold (Munn 1992: 96). Instead of the conception of the present as effect of the past, the present also shapes the past as not all pasts remain as a powerful force. Moreover, the present is catalytic of life experiments although the configuration of the results cannot be predicted (May 1996). Among various cultural groups, tensions, or lack of, in the coexistence of past and present can be bodily expressed. As an ordinary phenomenon, it can be culturally observed through the bodies of healers and their embodied spirits, which are moral voices, temporal markers and sources for restoring wellbeing (Lambek 1996; Masquelier 2001). As an extraordinary phenomenon, the legacies of experiences of serious violations persist in the form of powerful spirits that intrusively strike in people's bodies and disrupt their welfare and sense of security. The attack signals the tense coexistence of past and present and the need of accountability to attain resolution. Thus, embodied accountability practices in Africa (Igreja 2009) and around the world (Mueggler 2001 for south-west China; Kwon 2008 for Vietnam) exceed global accountability projects given that there are no predetermined timeframes over which an alleged serious violation ceases to require liability.

In spite of the greater visibility of global legal accountability, it does not capture the array of notions and experiences of serious violations and the diversity of sites and practices of accountability centred on the intrusive embodiment of aggrieved spirits (Allen 2006; Clarke 2009; Meier, Igreja and Steinforth 2013; Mueggler 2001). Various studies have revealed how, in different regions of Mozambique, war survivors were actively dealing with the legacies of the civil war (Bertelsen 2016; Honwana 2003; Igreja 2009;

Lubkemann 2008; West 2005; Wiegink 2013) in a national political context dominated by amnesty laws (Igreja 2015b). Yet none of these works has analysed ongoing struggles and negotiations of temporalities of accountability for serious unresolved offences. In Gorongosa, the Portuguese colonial occupation, anti-colonial war, civil war and famine contributed to the perpetration of serious violations against and amongst civilian populations. A proper understanding of the negotiations for unleashing accountability requires a consideration of at least two aspects.

First, an analysis of how the dynamics of 'fragmented war', whereby political violence, which is prepared and undertaken by national political actors, is also 'appropriated by local actors in service to their own agendas' (Lubkemann 2005: 493). The spiral of violence in such cases involves processes of victimisation perpetrated not only by soldiers against civilians but also among civilians themselves. In this regard, a focus on the masterminds as ruled by the global accountability mechanisms fails to grasp the array of actors and forms of victimisation that those involved regard as serious.

Second, from a cultural perspective the notion of serious offences is not confined to the visible, material and quantifiable dimensions as is the case of the Rome Statute (ICC 2002). In numerous African nations, human bloodshed and homicide pollutes the earth to the extent that whole communities can be endangered (Douglas 1980: 69). Following violent deaths, disturbances continue among the living regardless of the passage of time and number and identification of the corpses' location. Persistent bodily disturbances enhance the sense of collective insecurity, which can be halted through locally meaningful forms of resolution.

In this regard, similar to the analysis of social and political relationships proposed by Todd May (1996: 301) following Deleuze and Guattari, embodied accountability is conceived as processes of 'interaction among various disparate but often intersecting practices which are themselves evolving' through networks that sometimes mesh micropolitical and macropolitical structures. In Gorongosa, embodied accountability for serious violations unfolds in multiple sites and stretches over indeterminate time spans through a set of contradictory and overlapping actions of spirits, intergenerational family networks, healers particularly from AMETRAMO (Associação dos Médicos Tradicionais de Moçambique), judges of community courts, and macropolitical structures represented by state officials particularly from the police and the judiciary (Igreja and Racin 2013).

Methodologically, this study is based on long-term qualitative and quantitative research initiated in April 1997 in Gorongosa district, focusing on the analysis of war and famine experiences and extant resources to come to terms with a multiplicity of legacies. The specific data used for this analysis were gathered during fieldwork in 2014. Methods involved interviews with key participants and participant observation in local community courts and yards of healers that were catalytic for the resolution of the case under analysis.

Spiralling of political violence in Mozambique over time

The majority of colonial regimes exerted influence in the colonies through exploitative policies, Catholic religious groups and the colonial army and police. In former Portuguese colonies, the colonial authorities expanded their territorial presence in the beginning of the twentieth century by defeating local kingdoms and chiefs and gradually began imposing colonial order. The Portuguese colonial officers also understood the need to ritualise violence as a way to legitimise their mechanisms of coercion (Igreja and Skaar 2013). Thus, in the centre of Mozambique, people's memories of colonisation and some of the mechanisms of accountability have focused on certain violent actions of the Portuguese administrators and the *cipaios*. The latter were a group of African men that worked for the colonial administration as policemen. They violently coerced the local populations into paying the hut tax and forced them into labour camps (Pitcher 1998). However, the collaboration and complicity between *cipaios* and some of the community members, who betrayed their families and neighbours and contributed to increment violence in the communities, has received scant attention in studies of the legacies of colonialism in Mozambique.

Following nearly thirty years of effective Portuguese colonisation, the Mozambican Liberation Front (FRELIMO, Portuguese acronym) was founded in 1962 to wage an armed struggle for independence. The country reached independence in June 1975 at the height of the Cold War. However, the euphoric celebration of independence did not resolve the conflicts that were unleashed in the context of colonial occupation. Over time these unresolved conflicts culminated in a civil war involving the Frelimo government against the rebel movement Renamo (Cabrita 2000; Ncomo 2003). The white minority regimes in apartheid South Africa and former Rhodesia (now Zimbabwe) further fuelled the civil war in Mozambique.

In Gorongosa district, the war-zones were stratified between government and Renamo-controlled areas. The government army controlled the communal villages where many families lived closer to one another. The Renamo rebels controlled the *madembes* (old residences) where people lived in scattered homesteads. The sides in conflict fought for the control of the people in order to sustain their war efforts. The war also involved bloody battles and killings, initially between the Frelimo government and Renamo armies. In the mid-1980s, the Zimbabwean army became directly involved in battlefield fighting alongside the Frelimo government's army. Yet the violence escalated in such a way that family and community members, willingly or through coercion, also became involved in violence and committed serious violations among themselves (cf. Lubkemann 2005).

In the midst of the war, people still continued to work in the fields, get married and divorced, and carry on their business. Some sort of social life survived (Lubkemann 2008; Nordstrom 1997; Schafer 2007), although the devastating effects of the civil war permeated these social relations, as the case of Rita and her kin below attests. Communities became bitterly divided and these divisions fostered relations of mutual distrust and hatred, and lack of family and community solidarity and support, which also led to numerous acts of violence, detentions and torture, rape and the murder of kin and neighbours. Severe droughts and famine during 1987–1988 and 1990–1992 aggravated the wartime suffering and interpersonal violations (Igreja 2012b).

On 4 October 1992, the Frelimo government and Renamo signed the General Peace Accord (AGP, Portuguese acronym). The agreement was pivotal for setting the first multiparty elections and national democratic parliament. Yet the civil war left unresolved conflicts at the level of central political actors and local communities and families, which did not simply disappear with the attainment of the AGP and engagement in the democratisation project (Igreja 2013). Instead, the AGP created the context whereby not only the violations of the civil war but also of the colonial period were brought to the fore to be accounted for by the alleged perpetrators and their families. Before dwelling on the case of Rita, the cultural context and practices in Gorongosa that renders legitimate accountability for violations from different temporalities is presented.

Cultural notions of serious offences

Gorongosa is a rural district featuring political, legal and religious pluralism. Community and state institutions and actors diversely shape everyday life. The region is also characterised by patrilineal kinship, polygamy and monogamy within an agriculture-based economy. The recent introduction of new technologies such as cell phones, television and videos has expanded the local economy. People speak Chi-Gorongose, a mixture of various regional languages and there is, as well, a minority of Portuguese speakers. Social life in the district unfolds in an unpredictable environment marked by visible and invisible forces, which can paradoxically interact, protect, amplify or threaten and extinguish the human body (*manungo*). As for many cultural groups in sub-Saharan Africa, for the Gorongosas life and death are mere stages of life and death is politicised, as the effacement of the body becomes an extension of life (Clarke 2009: 28). Life beyond corporeal death is made possible, as stated above, through the interventions of spirits, which are believed to meddle in the realm of corporeal life.

Locally, people distinguish two types of traumatic death (*kufa*). One is the death of a person through disease (*kufa nanenda*) that other persons failed to care for (*ku koia*). The second is the death of a person through the breaking of taboos such as witchcraft and cannibalism (Igreja 2014b; cf. Geschiere 1997; West 2001) or by physical violence leading to spilling of blood (*shiropa*) and killing (*ku puthua*) without social legitimisation. The word *ku zunguza* has similar connotations to *ku puthua*, which means something very bad and dangerous that ends with a person's death. Both murder and death due to the co-occurrence of disease and neglect are regarded as serious violations, and serious consequences befall upon those regarded as bearing responsibility. It is believed that in cases of traumatic deaths, the spirit of the dead does *ku hirindzira* (vengeful return) to vindicate his death (Igreja 2010). Furthermore, in contrast to the notion of connecting the punishment for violent death to an individual perpetrator or mastermind (Hayner 2001), responsibility and guilt are also collective (Clarke 2009).

In Gorongosa, redress for serious offences is enacted through paying back (*ku lapiwa*). If the perpetrator fails to pay back, his/her extended kin members are doomed to suffer. This logic of embodiment of debt, moral fault and collective suffering (Honwana 2003: 72) amplifies the seriousness of the offence. The notion of serious violation is linked to the continuation of afflictions, which are believed to be sustained by invisible agents and mechanisms. Ongoing suffering is also maintained through silence, concealment (*ku fissa*), and denial (*upoca*). The relatives of the perpetrator are adversely impacted by wrongdoing in which they did not participate, and this is experienced as acts of unacceptable injustice that requires accountability. Given the patrilineal feature of this society, healers and judges often track the genealogy of serious hidden offences in the bodies of plaintiffs and his or her patrikin (*dzindza*), and the accountability process initiates when the elder person among the patrikin is present to do *ku himirira* (*de jure* responsibility).

Case study of embodied accountability

The case of Rita ignited in 2014 when she was around 25 years old. She started to noticeably shake her body, which was identified as a sign of the presence of harmful spirits, but, each time it happened, the spirits did not talk. She became very aggressive, which impacted on her relationship with her third husband and her paternal kin in general. The previous husbands had divorced her because she had stopped bearing children. For six years, since her first child was born in 2008, she had failed to become pregnant again. She started having difficulties in sleeping, experiencing nightmares almost every night, headaches, poor appetite, and, according to her family, Rita became visibly careless with her own safety and her child and was no longer working in the field to sustain herself. Rita asked her father Velado for help, but he initially refused to accompany her to the healers to examine the etiology of her predicament. When she subsequently went to the healer with her ex-husband, the healer identified the existence of a serious problem in her family and summoned her father. Yet again he gave an excuse to avoid attendance at the healer's house. In this case, Rita decided to report her case to the local community court. The court advised Rita to report to the police because her father had yet again snubbed the judge's orders. Rita reported her case to the police and the police officer in charge gave her a kind of subpoena that prodded her father (Velado) to attend the healer's request for treating his daughter, as otherwise he could be summoned to the police for negligence.

Velado bowed to the pressure and the whole family went to AMETRAMO to initiate the examination of the possible causes of Rita's predicament.

The healers in charge at AMETRAMO played the drums, Rita's body started to shake and matters reached a climax when, for the first time, she became possessed to the extent that the spirit in her body conveyed the willingness to talk about his return and the havoc he was presumably wreaking on Rita's body and her family.

Civil War Offences: Murder, Torture and Rape

The spirit identified himself as Carabino. The healer in charge asked Velado and his family if they knew Carabino. Initially Velado answered that he had never heard of such a person. The healer asked Carabino to present some clues to help Rita's family remember. The spirit refused while asserting that 'if you continue to do *ku fissa* (to hide) and *upoca* (deny responsibility) and pretend that you don't know me, I will drive this *txiquiro* (spirit's host, in this case Rita) crazy, kill her and then kill all of you one by one, as you did it to me and then raped my wife'. At that moment, Velado's wife, said, 'There was a person who used to be our neighbour in the communal village. One night, his wife and children were taken and they never came back.' Nevertheless, Velado's wife clarified: 'I just heard (*kuvba*) from other people, I did not see with my eyes. Someone can only say what they saw with the eyes.'

The healer said, 'This is *gamba*' and insisted, '*Dairane, micero ienei kuenda ku tale*,' which translated, 'Account yourselves, otherwise this problem is going far away.' In this understanding, the conflict is conceived as floating in time; the present is unable to redeem the past and such tense coexistence is problematic, given that over time the past also defines the future. The virtual future, or the future as a possibility, is the grim reality of Rita's failure to conceive children, thereby being deprived of participation in processes of social reproduction.

Following the healer's warning that the problem was growing out of proportion, Carabino, the spirit, also interjected, 'Velado why do you keep quiet? Is your mouth dry?' Velado's brother intervened. 'The only person in our family that knew many things that were happening in the communal village was Velado because he was a militia, and many times he was away from home at night.' The healer said that he would terminate the session because Velado and his family were not serious about helping their daughter Rita. He stopped the session and appealed to Velado and his kin to think seriously about the fate of Rita and their extended family.

A week later in response to Rita's deterioration, Velado requested a hearing session. The spirit of Carabino was called and again came out through Rita's body. Carabino said, 'Tell in front of everyone what you have done that made me return to afflict!' Velado started disclosing what had happened. As his brother had already hinted the previous week, Velado confirmed that he had been a government militia during the civil war and he was in charge of the security of the communal village where he lived with his family and many other local families from the region. He informed the audience, as his wife had alluded, that one of their neighbours in the communal village was a woman named Santalucia who lived with her two children. Santalucia and some other families in the communal village had been kidnapped from the Renamo-controlled area and brought as *recuperados* (recovered people) to settle in the government area. Santalucia's husband, named Carabino, was absent when his wife and two children were brought to the communal village. However, a few days later and by his own initiative, Carabino ventured into the communal village in search of his wife and children. He presented himself to the authorities; his wife was interrogated about whether she knew Carabino to which she answered positively. Carabino was released and allowed to settle with his family. A few days afterwards, at dawn, Velado and two other government militias went to Carabino's hut, seized him, his wife and two children. They were taken to a nearby bush area, Carabino's eyes were covered, and Velado executed him in situ. Yet Velado tried to defend himself, alleging that it was not of his own volition: 'the commandant of *matropa* [government soldiers] at the time ordered me and two other militias to execute Carabino because it was believed he was a *Matsangaissa* [Renamo soldier, or Renamo supporter]'. Carabino rebuked this defence by stating, 'You killed me not because of the commandant but because you wanted my wife.' Then he asked, 'Did your commandant also order you to do *ku cita* (rape) my wife?' Velado did not have a response.

Since the wife of the deceased was not present and could not be found, the healer in charge asked if the other two militias who had accompanied Velado during the murder could be located. Velado answered that one of them was dead but one named Colimba was still alive and living in a nearby village. The healer grabbed his cell phone while demanding that Velado provide the cell phone number of his former colleague. Colimba responded to the call and a day later he attended the forum in the healer's house. Initially, some people in the audience were sceptical of what Colimba could add to these confrontations, given that, as someone said, 'He is not a *nhampena* [impartial]' – he had been friends with Velado in the war. In this regard, the healer warned Colimba from the outset 'not to write in the water', which meant not to tell lies.

In response to the healer's question, Colimba revealed: 'I did not receive any orders directly from the commandant; it was Velado who recruited me to seize that man [Carabino].' He was also asked about the alleged rape of the deceased's wife. Colimba answered: 'It was Velado who raped her. My colleague and I were ordered to look after the children while Velado was raping the woman.' It was further revealed that, following the assassination and the rape, Velado ordered the widow and her children to move back to the Renamo-controlled area, alleging that, according to Colimba's testimony, 'you do not deserve to stay in the communal village because the meagre food we have is only for Frelimo members; you are Matsangaissa, and Matsangaissa alone will give you food'. Following these revelations, Velado loosened his stance of putting blame on others. He acknowledged that he had been consumed by *ku timisira* (greed) and *ntchange* (jealousy) as, when he first saw Santalucia, he thought that she was single and hastily expected to marry her as his third wife. Yet when her husband Carabino emerged from the bush, Velado felt *shungwe* (rage) instead of *ucisse* (compassion) for the *recuperados*.

Once the contours of the wrongdoing were established, the healer asked the spirit about the type of penalty required for the loss he had incurred. The spirit ordered '*tire kuda ku barirwa*', which translated, 'you must give birth to a child for the spirit'. The healer asked why the spirit was setting a heavy penalty on his perpetrator. The spirit answered 'because I also wanted to live, I also wanted to see my children grow up'. The healer responded, 'But the war is over; we now have reconciliation and forgiveness.' The spirt asserted, 'The war is not over for this family' and he warned '*aripo muno munango ada putiwa* [there is another person that was murdered].'

This information increased the sense of despair of Velado and his kin as they had thought that the resolution of the civil war-related conflict would put an end to their suffering. While people know that in general spirits are linked to tragic experiences of possible pasts, no one can tell in advance the temporality that the disturbing spirit encapsulates without the spirit's voice. In this regard, Velado and his family asked, 'Which other dead person? If you know something why not just tell us so that we can have a relief?' The spirit responded in a metaphorical way, 'Don't put the cart in front of the donkey. The person that was hurt by the offence will come by himself to report.' This suggested that the other spirit on the loop needed a forum to be heard too as precondition to eventually resolve this conflict.

Unresolved Serious Offences from the Colonial Context

The healer initiated a general song which is used to respectfully call the spirit to come out and reveal himself to the audience. This time it was Rita's sister, named Zabel, who first showed signs of being possessed by shaking her body in an uncontrolled manner, and once fully possessed, the spirit started singing his own song of *nhatua* (suffering) and loss,

> *Nhatua mai* [suffering mother]; *Inda luza* [I lost]; the ones that slept lost; I lost, the ones that slept lost; the one that sleeps loses; I am here because I lost; I lost; I lost; I lost; I lost; I have arrived; I am tired; *tchena moio mai* [I am murmuring all the time]; I am tired.

Yet besides the appealing function, it is known that songs also, 'either directly or implicitly, make symbolic statements about common experiences and concerns' that can be considered as popular consciousness and safe medium of protest (Vail and White 1983). Following the initial song, the spirit acting through the body of Zabel sang another song which gave some indications of the temporalities of the wrongdoing,

> The war of the Germans, there is a need to make an effort otherwise you die; there is war, the war of Caetano there is a need to make an effort [Marcelo Caetano was the last Prime Minister of the colonial Portuguese government. People in the centre of Mozambique use his surname as synonymous with the Portuguese colonial authorities]; there is war, the war of the Germans, you have to make an effort otherwise you die; there is war, there is the war of the Caetano there is a need to make an effort otherwise you die; today there is a war; there is war of the Germans, you have to make an effort otherwise you die.

The song alluded to the colonial wars unleashed by the Germans in East Africa during World War I and by the Portuguese army. The references to these temporalities perplexed everyone, given the dominance hitherto of *gamba* spirits that vindicated civil war offences. Nevertheless, the evocation of an unsettled conflict of the colonial era indicated the problematic coexistence of pasts even more greatly distanced with the present.

The singing stopped and the spirit disclosed his identity as Munengueia. The healer asked Rita's father and kin who were present in the audience if they knew Munengueia. They all answered negatively. The healer warned the audience not to do *ku nhepa* (tell lies) while further asking if someone in their extended family might have clues (*lusso*) who Munengueia was. The spirit intervened by asking 'don't you know me? Where is Raiene?' The healer demanded of the audience: '*dairane* [give an answer]!' Zabel's father, Velado, confirmed that Raiene was his father but he had stayed home as he was ill. The healer stopped the session and demanded that Raiene be brought to the hearing. This is a typical demand of *gamba* healers to probe the spirits' accusations because the living are also suspicious that spirits might be inaccurate.

Three days later, Raiene appeared in the healer's house and another hearing session was undertaken. The spirit manifested to the audience again through the body of Zabel and initially said, 'That which has horns cannot be wrapped up in a bag,' which is a local proverb that means 'there is nothing that can be done secretly without being unveiled'. Unlike his son Velado, Raiene did not attempt to deny knowledge of the identity of his accuser, Munengueia. He acknowledged that they were neighbours during the colonial period in a village in Gorongosa. At the time, the Portuguese colonial regime relied on the *cipaios* (mentioned above) to enforce the hut tax policy and forced labour as a punishment. The spirit of Munengueia revealed that Raiene betrayed him by reporting his whereabouts to the *cipaio* in exchange for money. During the persecution, Raiene actively helped the *cipaio* and when Munengueia was seized, he was beaten hard around the head. He fainted, and after a while he recovered consciousness, but died some hours later, presumably from an internal injury in the head. For Munengueia, Raine was partly responsible for this tragedy as he had actively assisted the *cipaio* in his violent capture and subsequent death.

The mentioning of the *cipaios* sparked a process of collective remembering through the evocation of some names of former *cipaios* that had operated in Gorongosa. This evocation was part of the circumstantial attempts to concretise the context and events under consideration. The healer in charge and some other elders in the audience probed for some names. The healer said, 'There was a *cipaio* named Totcho; when he found people, he used to show no respect. After threatening people, he used to say that no one could do anything against him. Is this the one?' Raiene answered negatively. An elder in the audience asked, 'Was it Mutsumbandaita; once a transgressor at the time saw Mutsumbandaita in the neighbourhood, the transgressor instantly regretted a wrongdoing because he had no mercy.' Again Raiene answered no. The healer probed with another name: 'It was not Tungadza? He was very bad; when he arrived in a household no one managed to get close to him because of fear.' While drawing laughter from the audience, Raiene answered positively.

Once these broader contextual aspects were disclosed – which also had historical relevance in terms of analysis of the diverse imaginaries of violence crystalised in names and the type of local relations among colonial administrators, *cipaios* and various local people – Raiene admitted that he had collaborated with Tungadza. He helped him capture insubordinate young men in the region. However, Raiene claimed that he had not seen Munengueia dying because 'immediately following his capture, the *cipaio* took Munengueia to Vila Paiva'. This was an attempt to deny responsibility for the murder by suggesting that he had handed Munengueia over alive; therefore, he had betrayed but not killed the accuser. The spirit rebuked this defence by alleging that, following his death, Raeine hastily moved from the region to Chimoio (a neighbouring town) to avoid local accountability and buy time to instil forgetfulness. On this account, too, Raiene denied that he ran away to avoid accountability. He alleged that together with his family, they fled the region because of the intensification of the Frelimo-led anti-colonial war against the Portuguese army in 1973. In the meantime, Raiene's family indicated that they knew the patriarch had a dubious character, but they were not aware that he had participated in actions that led to a neighbour's death during the colonial temporality. To conclude, the healer asked the spirit what he wanted as compensation for that tragic loss. The spirit said that he wanted to have a *txiquiro* [body vessel] to work as a healer. Yet the healer opposed that request by arguing that there were still doubts regarding whether Raiene could be directly blamed for his death. The spirit argued that if Raiene had not betrayed him he could still be alive like many other young men that escaped unjust punishment by seeking refuge in colonial Zimbabwe. The healer asserted that if Munengueia was right in his claims, he should doggedly seek vindication among the offspring of Tungadza and that, in the event of powerful signs, the case would be brought again to the healer. For the time being, the healer demanded that Raiene and his family issue an apology to Munengueia following the local cultural norms.

Embodied accountability and temporalities of resolution

The developments of formal legal mechanisms of accountability for serious violations perpetrated in contexts of mass political violence have boosted the practice of witnessing in the past five decades. The idea of era of the witness consists in the incremental significance of victims' voices to present evidence of serious violations for the purpose of criminal accountability (Wieviorka 2006). Yet these international efforts and expectations on matters of legal accountability have not seriously taken into account the plurality of accountability sites and non-legal practices prevailing in numerous communities torn by violent political conflict in Africa and around the world.

As the case of Rita illustrated, experiences of violence, serious breaches in interpersonal relations and enduring illnesses are conceived as serious injustices scripted in people's bodies, and the consequences are not bound by strict temporalities and individual responsibility. Instead, embodied accountability consists of ongoing individual (bodily) and collective (social) disturbances which triggers unpredictable struggles and negotiations to set the temporalities and culprits of serious offences within kin circles.

Spirits and intrusive spirit possession signal ruptured temporalities, and these agents pinch alleged perpetrators and their offspring in ways that unsettle their everyday lives and reinforce the perception that the past as they know it, remember and have been told (breaks of order, suffering) is unending. The spirits in this case – of Carabino and Munengueia – constituted traces of serious wrongdoings perpetrated in the temporal context of the civil war and colonial occupation in Mozambique. Unlike the official mechanisms of accountability founded upon temporal and hierarchical models of organisation, the accountability case under analysis here did not suggest a hierarchy of importance, even if the initial case referred to the recent past and the subsequent case was related to a more distant past. Additionally, contrary to notions of individual accountability in cases of serious violations, the Gorongosas emphasise collective responsibility. In this regard, the kin of the alleged perpetrator are not immune from the retributions of the spirits. When the spirits strike the kin of the alleged perpetrator, they experience these disturbances as injustices. They stridently argue that they are suffering as a payback for a serious wrongdoing in which they had no part. The effect of these collective retributions, however, as observed elsewhere, is to exert incremental pressure upon the alleged perpetrator to give an account of the past evil deeds (cf. Douglas 1980: 69).

When the pressure from within kin circles is not powerful enough to expand accountability actions, accountability incrementally evolves in contradictory ways to include macropolitical structures represented by the state agents. It is contradictory, given that, statutorily, the Mozambican state claims to be ruled by secular values and strict definitions of temporality to reckon with the past, and precludes involvement of spirits on matters of the state. Yet, in everyday life, numerous state agents operate according to indigenous forms of power and authority founded upon local beliefs and fears of spirits (Igreja 2014b). In this case, state officials in local police stations are conscious of the impact of harmful spirits in people's lives and they collaborate in their resolution by initially writing a kind of subpoena, which often has the effect of pressing the refusing side in the conflict to participate in search for a resolution. In spite of their involvement, the police officers are not specifically aware that the accountability they help to accelerate is related to serious violations that unfolded during colonial and the civil war periods. Accountability for such types of cases were declared invalid, at least officially, through political forgiveness and amnesty laws (Igreja 2015b). Yet, for the Gorongosa inhabitants, since the volition and puissance of aggrieved spirits escapes their immediate control, the state amnesty laws to putatively undo past serious violations have little performative effects. The local litigants contravene existing amnesty laws by presenting their claims in the cultural idiom of bodily suffering wreaked by supernatural forces. This idiom of accountability and litigation works to pre-empt state agents from evoking official laws to put a lid on conflicts and claims related to past political violence. Ultimately, state officials indirectly help in consolidating embodied accountability that can unpredictably expand the temporalities of serious wrongdoings and the healing techniques required to heal some of the legacies of colonialism and civil war in Mozambique.

Conclusion

When the alleged serious wrongs perpetrated during colonial and civil war temporalities were publicly heard and acknowledged and the penalties and issuing of apologies were accepted by Rita's paternal kin, the process of resolution gained a new impetus. The serious wrongs from the civil war were addressed by offering one young girl to the aggrieved spirit. The girl became the receptacle of the spirit, therefore she should not be subjected to any form of abuse and must be treated with utmost respect (*makodzo*). In turn, the spirit committed himself to maintaining and reinforcing peace and security in the extended family. To actualise their mutual commitments, Rita's kin perform occasionally ceremonies to worship the spirit of Carabino, and when the hosts and the family are bothered by misfortunes they initially consult his spirit to determine what might be happening. In this case, the coexistence of past and present is no longer problematic because the successful resolution of grievances allows Rita and her kin to participate in processes of social reproduction.

Despite the flexibility of embodied accountability, the resolution is still not instantaneous. In both claims, there was denial, intense deliberations, postponements and requests for further evidence that met local standards of truth. While the procedures in both cases were similar, the outcome differed in that, in the case related to the violation in the colonial period, the healer could not establish a direct link between Raiene's betrayal and the murder of

Munengueia. The healer instead asked Munengueia for more time and more pressure exerted directly among the offspring of the former *cipaio* and now deceased Tungadza. Regrettably the beginning of new military confrontations between the Frelimo government's army and Renamo did not allow me to conduct follow-ups to further determine the struggles for accountability that might have arisen in the case of Tungadza offspring.

Nevertheless, the absence of global accountability actions for the civil war violations in Mozambique, the refusal of the state authorities to pursue accountability at the end of the civil war in 1992 and the weakness of formal civil society organisations to exert pressure on matters of accountability evidence the vitality and significance of embodied accountability to shun impunity and contribute to justice in local cultural terms. Additionally, the involvement of supernatural agents as sources of conflict and resolution bespeaks the closer affinities between justice and healing of bodily afflictions to the extent that the lack of justice results in disruption of interpersonal relations and outbreak of serious illness among the people involved. The deliberations and probing of the causes also create a context whereby memories of diverse temporalities are evoked, transforming the resolution forum into a process of bearing witness, remembering, social protest and both reinforcement and change of collective identities.

References

Allen, Tim. 2006. *Trial Justice*. London: Zed Books.

Al-Saji, Alia. 2004. The Memory of Another Past: Bergson, Deleuze and a New Theory of Time. *Continental Philosophy Review* 37: 203–239.

Bertelsen, Bjørn. 2016. *Violent Becomings*. New York: Berghahn Books.

Cabrita, João. 2000. *Mozambique: The Tortuous Road to Democracy*. New York: Palgrave.

Clarke, Kamari. 2009. *Fictions of Justice*. Cambridge: Cambridge University Press.

Deleuze, Gilles. 1989. *Cinema 2*. Trans. H. Tomlinson and R. Galeta. Minneapolis: University of Minnesota Press.

Deleuze, Gilles. 1991. *Bergsonism*. Translated by Hugh Tomlinson and Barbara Habberiam. New York: Zone Books.

Douglas, Mary. 1980. *Evans Pritchard*. Brighton: The Harvester Press.

Geschiere, Peter. 1997. *The Modernity of Witchcraft*. Charlottesville: University of Virginia Press.

Hagan, John, Heather Schoenfeld and Alberto Palloni. 2006. The Science of Human Rights, War Crimes and Humanitarian Emergencies. *Annual Review of Sociology* 32: 329–349.

Hayner, Pricilla. 2001. *Unspeakable Truths*. New York: Routledge.

Honwana, Alcinda. 2003. Undying Past. In Birgit Meyer and Peter Pels, eds., *Magic and Modernity*, 60–80. Stanford, CA: Stanford University Press.

Igreja, Victor. 2008. Memories as Weapons. *JSAS* 34: 539–556.

Igreja, Victor. 2009. Justice and Reconciliation in the Aftermath of the Civil War in Gorongosa, Mozambique Central. In Kai Ambos, Judith Large and Marieke Wierda (eds) *Building a Future on Peace and Justice*, 423–437. Berlin: Springer.

Igreja, Victor. 2010. Traditional Courts and the Struggle against State Impunity for Civil Wartime Offences in Mozambique. *Journal of African Law* 54: 51–73

Igreja, Victor. 2012a. Multiple Temporalities in Indigenous Justice and Healing Practices in Mozambique. *IJTJ* 6: 404–422

Igreja, Victor. 2012b. Negotiating Order in Postwar Mozambique. In Helene Kyed, João Coelho, Amélia Souto andSara Araújo, eds., *The Dynamics of Legal Pluralism in Mozambique*, 148–166. Maputo: Kapicua.

Igreja, Victor. 2013. Politics of Memory, Decentralization and Recentralization in Mozambique. *JSAS* 39: 313–335.

Igreja, Victor. 2014a. Legacies of War, Healing, Justice and Social Transformation in Mozambique. In Brandon Hamber andElizabeth Gallagher, eds., *Peacebuilding from a Psychosocial Perspective*, 223–254. Springer: New York.

Igreja, Victor. 2014b. Memories of Violence, Cultural Transformations of Cannibals, and Indigenous State-Building in Post-Conflict Mozambique. *CSSH* 56: 774–882.

Igreja, Victor. 2015a. Media and Legacies of War. *Current Anthropology* 56: 678–700.

Igreja, Victor. 2015b. Amnesty Law, Political Struggles for Legitimacy and Violence in Mozambique. *IJTJ* 9: 239–258.

Igreja, Victor and Elin Skaar. 2013. A Conflict Does Not Rot. *Nordic Journal of Human Rights* 31: 149–175.

Igreja, Victor and Limore Racin. 2013. The Politics of Spirits, Justice, and Social Transformation in Mozambique. In Barbara Meier andArne Steinforth (eds), *Spirits in Politics*, 181–204. Frankfurt: Campus Verlag.

International Criminal Court (ICC). 1998. Rome Statute of the International Criminal Court. The Hague.

Kwon, Heonik. 2008. *Ghosts of War in Vietnam*. Cambridge: Cambridge University Press.

Last, Murray. 2000. Children and the Experience of Violence. *Africa* 70: 359–393.

Leccardi, Carmen. 1996. Rethinking Social Time: Feminist Perspectives. *Time and Society* 5: 169–186.

Lambek, Michael. 1993. *Knowledge and Practice in Mayotte: Local Discourse of Islam, Sorcery, and Spirit Possession*. Toronto: University of Toronto Press.

Lubkemann, Stephen. 2005. Migratory Coping in Wartime Mozambique. *Journal of Peace Research* 42: 493–508.

Lubkemann, Stephen. 2008. *Culture in Chaos*. Chicago, IL: University of Chicago Press.

Masquelier, Adeline. 2001. *Prayer Has Spoiled Everything*. Durham, NC: Duke University Press.

May, Todd. 1996. Gilles Deleuze and the Politics of Time. *Man and World* 29: 293–304.

Meier, Barbara, Victor Igreja and Arne Steinforth. 2013. Power and Healing in African Politics. In B. Meier and A. Steinforth (eds), *Spirits in Politics*, 15–36. Frankfurt: Campus Verlag.

Mueggler, Erik. 2001. *The Age of Wild Ghosts*. Berkeley: University of California Press.

Mueller-Hirth, Natascha. 2017. Temporalities of Victimhood: Time in the Study of Postconflict Societies. *Sociological Forum* 32(1): 186–206.

Munn, Nancy. 1992. The Cultural Anthropology of Time. *Annual Review of Anthropology* 21: 93–123.

Ncomo, Barnabé. 2003. *Uria Simango*. Maputo: Edições Novafrica.

Nordstrom, Carolyn. 1997. *A Different Kind of War Story*. Philadelphia: Pennsylvania University Press.

Pitcher, Anne. 1998. Disruption without Transformation. *Journal of Southern African Studies* 24: 115–140.

Posner, Eric. 2009. *The Perils of Global Legalism*. Chicago, IL: University of Chicago Press.

Schafer, Jessica. 2007. *Soldiers at Peace*. New York: Palgrave Macmillan 2007.

Vail, Leroy and Landeg White. 1983. Forms of Resistance: Songs and Perceptions of Power in Colonial Mozambique. *The American Historical Review* 88: 883–919.

West, Harry. 2005. *Kupilikula: Governance and the Invisible Realm in Mozambique.* Chicago, IL: University of Chicago Press.

Wiegink, Nikkie. 2013. Why Did the Soldiers Not Go Home? Demobilized Combatants, Family Life, and Witchcraft in Postwar Mozambique. *Anthropological Quarterly* 86: 107–132.

Wieviorka, Annette. 2006. *The Era of the Witness.* New York: Cornell University.

7 Still waiting

Victim policies, social change and fixed liminality

Natascha Mueller-Hirth

Introduction

Waiting is a part of everyday life. We all wait, whether it is for a bus to come or the rain to go, for a Yes or a No, or for our hair to grow, as Dr Seuss readers will remember. In the children's author's beautiful and inspiring journey through life's triumphs and trials *Oh, the Places You'll Go*, we briefly pause in the Waiting Place. We arrive here in a state of confusion ('you can get so confused') and travel through 'weirdish wild space'. The Waiting Place is 'a most useless place'. Everyone here is 'just waiting'; we see anxious characters sitting on a sofa, just passing time; others have formed a long orderly queue in front of a door that leads nowhere. The waiting place is not a happy place. However, as we turn the page, we find that this place is 'not for you' and that 'somehow you'll escape all that waiting and staying. You'll find the bright places' (Dr Seuss 1990).

In my research with victims and survivors of gross human rights violations of the Apartheid era in South Africa, people were very often talking about waiting. For example, S'bu explained that 'we are still waiting, I don't know for how long'. The temporal construct of 'still waiting', as it appears in this sentence, was used many times when participants spoke about reparations for the past and articulated their hopes for transformative change in the future. It is also noteworthy, in S'bu's words, that the duration of the wait is unclear: 'I don't know for how long'. To be 'still waiting' is one of a range of temporal metaphors participants used to describe their lives in post-Apartheid South Africa, such as 'still struggling', 'being left behind', or 'moving on'. Becoming aware of these temporal metaphors led me to engage with the scholarship on time and temporality to better understand how people's past and present experiences of violence impact on their being in time, and why some victims felt out of sync with the dominant forms of temporality (Mueller-Hirth 2017 develops an analysis of temporalities of victimhood).

In this chapter, however, I specifically examine the experiences and effects of waiting for victim support, recognition and social change among those affected by Apartheid-era human rights violations, 20 years after the democratic transition. South Africa is often regarded as an example of a society that has successfully dealt with its violent past. The country's approach to dealing with the legacy of 50 years of legislated racial oppression famously involved a truth commission that provided for conditional amnesty and has become a model for dealing with the past for other post-conflict settings (Chapman and Van der Merwe 2008). Transitional justice strategies also included reparations and rehabilitation, while affirmative action policies sought to deal with the economic legacies of Apartheid. Nonetheless, South Africa remains a deeply unequal society that faces huge developmental challenges. Moreover, the government never fully implemented the TRC's reparations policy, leaving many victims disillusioned and disempowered.

The anthropological concept of liminality, as the middle stage of rituals, has been applied to examining moments of transition, but it might also be useful in understanding the enduring wait and senses of being left behind that participants in this study reported. I argue that waiting for delayed payments of reparations and for change more broadly has fixed some victims in a permanent in-between state. It is precisely the need to have to wait that leaves them unable to move on from their victimhood. To be the person that waits – be it for transformative change or for recognition of one's contribution in the past, like so many in this study – means to be powerless and in a state of uncertainty. Drawing on the literature on time and power, and on recent work on the politics of waiting in particular, we can see that waiting is not neutral: to enforce a waiting period, such as delays in the payment of special pensions to victims, or to raise false hopes, for example with regards to community reparations, is to exert power. As a strategy of power, waiting reflects and exacerbates inequalities. This became apparent in relation to victims as a marginalised group in general, and victims from rural areas in particular. One important caveat needs to be made before continuing: by focusing here on the temporal marginalisation of victims, I do not deny their socio-economic marginalisation. On the contrary, temporal domination is a key dimension of victims' marginalisation and is intrinsically linked to their multiple experiences of exclusion (Mueller-Hirth 2017).

This study was part of a larger research programme, a six-year Leverhulme Trust-funded project entitled 'Compromise after Conflict' that focused on the development of compromise amongst victims of conflict in Northern Ireland, South Africa and Sri Lanka.[1] The South African study included 38 in-depth interviews with victims/survivors of Apartheid-era gross human rights violations that were conducted during four months of fieldwork in 2011. All interviewees had suffered gross human rights violations as defined by South Africa's Truth and Reconciliation Commission (TRC), as well as experiencing the structural, legislated-for, violence of Apartheid. All interviews were conducted in the Western Cape Province, in a mixture of urban (Cape Town and Cape Flats townships; 25 interviews) and semi-urban or rural (Oudtshoorn, George, Mossel Bay; 13 interviews) settings, and in most cases at participants' homes.[2] Eight participants were female, 19 historically defined black, 18 historically defined coloured and one historically defined white. Interviews were semi-structured, addressing themes such as victim identity, compensation, justice, reconciliation, present socio-economic conditions and hopes for the future.

There are debates around the use of the problematic term 'victim', with much of the transitional justice literature employing the term 'survivor' instead when referring to people that have been affected by violence. My interview schedule included asking participants how they described themselves in relation to the violence they underwent; only about one-third called themselves 'victim' and even fewer preferred the term 'survivor'. Other terms participants used were 'freedom fighter', 'victor' or 'activist-survivor'. As I have explored more fully elsewhere, it is significant that the term 'victim' was most commonly used by participants in relation to their life circumstances and experiences in the present, and not in relation to violence in the past (Mueller-Hirth 2017). For the purposes of this chapter, I use the term 'victims/survivors' when I refer to research participants, but refer to 'victims' when referencing official victim policies, echoing the terminology of the TRC.

This chapter proceeds with an overview of the relevant literature on social time and waiting as an effect of power in the subsequent section. The second section examines participants' experiences of waiting. Section three explores the effects of waiting, which ranged from immobility to refusal. The fourth part of the chapter develops the notion of fixed liminality, which serves to describe the supposedly temporary but actually permanent state of victimhood some of the participants in this study felt as a result of the lack of transformative change in post-Apartheid South Africa.

Social time and temporal domination

Although this chapter focuses on the experiences of waiting in a post-conflict context, waiting is only one facet of temporal control and domination. Scholars of social time have demonstrated a range of ways in which time becomes the object of social control and temporal power relations are shaped, for example in relation to gender (Bryson 2007, Lahad 2012), sexuality and sexual difference (Halberstam 2005), employment and illness (Charmaz 1991) and how social events reflect particular dominant temporal expectations (Zerubavel 1981). As has been argued in the introduction to this collection however, to date, there has been little critical engagement with time in the analysis of issues pertaining to transitional and post-conflict societies (but see Igreja, 2012 on multiple temporalities in negotiating transitional justice; Mueller-Hirth 2017 on temporalities of victimhood; Langer and Brown 2016 on the order of events in peace processes).

It is then perhaps not surprising that the question of waiting in transitional and post-conflict contexts has not been examined to date – waiting is a less noticeable form of power. Auyero calls this the 'invisible tentacles': 'less obvious or violent forms of power that achieve the subordination of poor people by making them "sit and wait"' (2012: 15). Waiting is usually seen as wasted or unproductive time (Lahad 2012), which is in part because temporal hierarchies render speed privileged over slow processes: fast is synonymous with profitable efficiency, whereas slow means inefficiency and backwardness (Adam 2004). Barry Schwartz (1974: 867) first drew attention to the fact that 'the distribution of waiting time coincides with the distribution of power.' To be kept waiting is a social assertion that one's time and social worth are less valuable, whereas wealthier people have 'relative immunity' from waiting (ibid.). Waiting is one of the ways in which the link between time and power can be experienced; it 'implies submission' and it modifies the behaviours of those who are waiting for the duration of the wait (Bourdieu 2000: 228).

Importantly, those that wait do not necessarily know how long they will wait, which directs attention to two issues that were particularly significant themes in the interviews for this study: hope and uncertainty. As Bourdieu put it, 'delaying without destroying hope, adjourning without totally disappointing [...] is an integral part of the exercise of power' (2000: 228). On the one hand, temporal uncertainty can work as a form of social control (ibid.); on the other hand, waiting is itself characterised by a lack of accurate and reliable information (Auyero 2012). This has been demonstrated by a range of ethnographies on the 'politics of waiting'. In Auyero's (2012) study of waiting in a welfare office in Argentina, welfare clients described its effects as insecurity, powerlessness and a lack of agency. Harms (2013) describes how waiting caused uncertainty for residents anticipating eviction from a development site in Ho Chi Minh City. This uncertainty undermined their ability to organise their social relationships and to direct their use of time towards productive income earning strategies. The literature on the politics of waiting also serves to show that making (subordinate) people wait is a strategy of temporal domination by the state. Reid's (2013) work on surviving Hurricane Katrina is important in framing the data in this chapter as it highlights that social inequalities are exacerbated by waiting. Many of the participants in her study had few resources and waited for months to get assistance or spent hours on the phone to get more information, leaving them 'stressed, worried, and unable to feel stable or make longer term plans' (Reid 2013: 758). Conversely, because they assume a middle-class habitus, assistance programmes did not have negative impacts on higher-income nuclear family households.

In these studies, waiting is clearly shown to be an oppressive practice that has negative impacts on the emotional and material well-being of people. Although waiting is usually understood as dead or wasted time, it also has productive effects. Some manage to transform waiting into 'an economically productive and surprisingly empowering form of social experience' as Harms (2013: 347) finds in relation to the Ho Chi Minh City residents mentioned above. Waiting was oppressive for many of the residents, but some fought 'waiting with waiting and respond[ed] to enforced uncertainty with temporal nonchalance and indifference' (ibid: 360). Similarly, some participants in my South African study were refusing to wait any longer for the government to recognise their contribution to the liberation struggle and had started their own localised memorialisation initiative, as I outline below.

Much happens when people are forced to wait in queues and waiting rooms, too: people learn about resources and they create networks and friendships, making waiting 'doubly relational' (Auyero 2012: 72). Auyero (ibid.) demonstrates that waiting can also be an active, intentional and sometimes strategic process. This is borne out by other research, for example Gray's analysis of the experiences of female non-migrants during the 1950s emigration boom in Ireland, where waiting shaped women's everyday lives by affording them strategic opportunities for family and social networks. Waiting can also be productive, in the Foucauldian sense, because 'the urban poor, in the frequent encounters with politicians, bureaucrats and officials, learn to be patients of the state [...] Habitual exposure to long delays moulds a particularly submissive set of dispositions' (Auyero 2012: 9). This function of waiting draws attention to the need for consent and compliance by those who wait, as Bourdieu describes:

> The games with time which can be played wherever there is power [...] can only be set up with the (extorted) complicity of the victim and his investment in the game. A person can be durably 'held' (so that he can be made to wait, hope, etc.) only to the extent that he is invested in the game to that the complicity of his dispositions can in a sense be counted on.
>
> (Bourdieu 2000: 230–31)

The reasons why people would consent to waiting are, as already indicated above, tied up with social status and power and waiting is therefore often designated to the poor and marginalised – those who literally cannot afford not to wait.

In summary, existing scholarship shows that waiting, as one form of temporal domination, is a strategy of power that can exacerbate social inequalities and affects those without resources disproportionally more severely. These insights have not been applied to examine the experiences of waiting in transitional and post-conflict societies, despite the fact that victim support policies are frequently delayed and waiting characterises the everyday lives of many victims/survivors. In the following section, I employ this literature to respondents' senses of 'still waiting'.

Experiences of waiting

In this section, I summarise narratives of waiting amongst victims/survivors in this study, which were most typically linked to compensation or reparations, to recognition and to transformative change that they fought and suffered for more broadly. Overall, waiting was perceived as exhausting and frustrating:

> People get tired. People don't want to go on anymore. And nothing happens, it is exhausting. [...]. And what I found out is, and that is why I always

> refer to rich and poor, if you are rich you have access to anything and everything in this world. But if you are poor you don't have access to anything.
>
> (Luisa)

As in this extract from my conversation with Luisa, who has suffered a continuum of direct and structural violence during and after Apartheid, there was a sense, for many participants, that waiting time is not distributed equally. Victims/survivors perceived that their own time was less valuable than that of other, more privileged people – including those who impose the wait.

Many victims/ survivors understood their long wait – be it for reparations or for social change more broadly – as a betrayal. This was sometimes directed at the government, for failing to fulfil the promises of the liberation struggle, or at particular individuals in the African National Congress (ANC) for 'the obscene wealth' (Eldric) they amassed. However, there were also a small number of participants that saw the pace of change as appropriate, given the challenges South Africa faces: 'I don't believe in immediate change, with something that has been entrenched for so long. And in fact some things persist to the present day' (Matthew). Participants such as Faisal were able to justify even the long wait of two decades: 'If you listen to what the government is saying, they are only saying that these things don't happen so fast. They are trying their best, the numbers which they say, in 2030 there will be so many jobs.' This kind of ongoing hope might be understood in terms of the great loyalty many South Africans continue to have for the ANC, the liberation movement turned ruling party that they fought in and made huge sacrifices for. Even participants who were critical of the ANC demonstrated this loyalty: 'People are angry with the ANC. They have also made a lot of mistakes. But people are saying sometimes, if your father is making you angry, would you leave your father? [laughs] You cannot leave your father. The ANC is our home. We need to look to how we can change it. Change it from the inside' (Tsepo).

Waiting for reparations

Perhaps the clearest illustration of unequal waiting times is provided by the reparations policy. South Africa is one of the best known cases of a country that has implemented restorative transitional justice mechanisms after political violence. The TRC began work in 1996 with a very wide remit: it set out to investigate any gross violations of human rights between 1960 and 1994 (the period from the Sharpeville massacre, where 69 peaceful protesters were killed by Apartheid police, to Nelson Mandela's inauguration as first democratic president). Gross human rights violations were defined by the TRC as 'the violation of human rights through the killing, abduction, torture or severe ill-treatment of any person emanating from the conflicts of the past and carried out or planned by any person acting with a political motive' (Truth and

Reconciliation Commission of South Africa, 1998: Vol. 5: 10). From 20,000 submissions, it classified 29,000 victims and documented 47,000 gross violations of human rights. It was also tasked with developing a reparation and rehabilitation policy. A President's Fund was established to provide:

- A reparations grant of R30,000 to those classified as victims (some victims also received urgent interim reparation of approximately R2,000)
- Symbolic reparation (national days of remembrance and reconciliation, erection of memorials and monuments, renaming of streets) and programmes to promote records of history, culture and art forms
- Medical and other forms of social assistance programmes to provide for medical benefits, education assistance and the provision of housing to address individual victims' needs
- Whole community rehabilitation programmes, aimed at promoting healing through services and activities, including health and education (Truth and Reconciliation Commission of South Africa 1999; Department of Justice and Constitutional Development 2016).

However, the Mbeki government decided that no reparations grants could be paid before the end of the Amnesty Committee, in 2003.[3] Officially recognised victims were therefore forced to wait for the reparations grant for up to seven and a half years after giving their testimonies (between 1996 and 1998). By contrast, perpetrators benefited from the TRC's amnesty process as soon as their case had been heard and amnesty granted. It is generally acknowledged that the amnesty deal was one of the central compromises of the transition; TRC Chairperson Archbishop Desmond Tutu has said that without the compromises made during the negotiations to ensure majority rule in South Africa, the country would have gone up in flames. Nonetheless, since amnesties were immediate, reparation should arguably have been immediate as well. The unequal waiting time can be seen to reflect an increasing political marginalisation of victims' concerns and needs after the transition. There continue to be regular protests by victims campaigning for a full implementation of the TRC's recommendations and a re-opening of the registration process, mainly associated with the Khulumani Victim Support Group. Funding of support for people affected by Apartheid-era violence has almost entirely vanished, however, because there is an understanding in the political and public spheres that the past has been sufficiently dealt with and that victims should long have moved on (Mueller-Hirth 2017).

Victims were not only forced to wait for the individual reparations grant, however. Other aspects of the reparations and rehabilitation policy, such as community reparations, are still outstanding, over two decades after the TRC began its work. When it emerged in 2013 that the President's Reparations Fund held over R1.1 billion (about £64 million) – due to interest, this was more than it did than before paying out individual compensation for victims in 2003 (Gontsana 2013) – some victims hoped that their long wait had not been in vain. Indeed, a report (South African Coalition for Transnational Justice 2012) identified several severely harmed communities where, according to government officials, reparative work is still meant to be implemented, including exhumations and reburials, community reparations and social services provision. In late 2013, the Department of Justice announced proposals that the money in the President's Fund should be paid out to fund infrastructure projects in 18 townships across the nine provinces, without clarifying why other TRC-designating communities were not chosen or re-evaluating survivors' needs. By 2016, further needs analyses had been conducted in Tumahole, Thabong (Free State), Paballelo, Seoding (Northern Cape), Lulekani and Ga-Matlala (Limpopo). The Reparations Fund, at the time of writing, holds about R1.9 billion (about £110 million) (Khulumani Support Group 2017).

This brings to mind Bourdieu's words, cited above, that 'delaying without destroying hope, adjourning without totally disappointing [...] is an integral part of the exercise of power' (2000: 288) – reparations have been delayed for over 20 years, but government has *not* stated that they will *never* be paid out. In marginalised and deprived communities where there are few employment opportunities and a lack of social, medical and educational services, the prospect – however unlikely – of some funds coming in is persuasive. The delayed payments of reparations in South Africa appear as a form of temporal domination: victims – who have mostly remained poor and have little political leverage – are made to wait for compensation. Conversely, those who are more privileged have been able to move on, or were able to refuse to wait. For example, having access to resources allowed other victims/survivors to better deal with the past, be it through counselling after the transition, financial resources or a clear break with the past. What is more, politicians such as former president Thabo Mbeki responded to victims' complaints about the delay in reparations by reminding them that the struggle was not about money, which served to delegitimise victims' needs and further marginalise them.

Special pension

At the time of research, a number of the interviewees were also waiting to receive their special pension or to be informed about whether they would qualify. The special pensions scheme, often referred to as the 'struggle pension' in common parlance, was established for veterans (i.e. people who served as soldiers in the liberation movements' armed wings such as the MK and APLA) and anyone else who, as member of a banned organisation, 'contributed for South Africa to become a non-racial constitutional democratic order' (Special Pensions Act 1996). Its purpose was to compensate those who, due to their involvement in the liberation struggle, lost the opportunity to provide for a pension (for a period of at least five years), and their surviving spouses and dependants. The scheme can also include non-military activists from civil society organisations, which characterised the Apartheid-era activities of many of the interviewees in this research.

The struggle pension scheme was marred both by fraudulent claims and by bona fide claimants' difficulties to access the pension (Faull 2011; Parliamentary Monitoring Group 2012). Applicants frequently had to wait years after having their application approved to receive their first pension payment. This can be illustrated with the story of Sipho, a 66-year-old man from Mossel Bay. He was exposed to beatings and harassment by the police, became a community activist with the United Democratic Front and spent several years in prison, where he also experienced torture. He had still not been paid and told me:

> They approved my pension fund in 2004. But the serious of the story, now 2011, they wrote me a letter, six years later. And about a month and a half ago, the lady that was working with this pension fund, I think she misused the funds, and now the money is gone. So I phone Pretoria again, they told me a long story, that maybe 2013 or 2014.

Others were told they needed to wait for verification of their claim. Participants often called many times to get further information, but were uncertain they would ever hear back: 'I applied for the special pension, but I haven't heard yet. Apparently, it is quite a mess. Maybe we will hear in the future' (Frans). As has been documented in the scholarship on temporal domination, waiting 'feels worse' when the waiter does not know how long they have to wait: 'punitive sanctioning through the imposition of waiting is met in the most extreme forms when a person is not only kept waiting but is also kept ignorant as to how they must wait' (Schwartz 1975, cited in Auyero 2012: 79). In the following extract from my conversation with Mpho, feelings of uncertainty – about waiting times but also about paperwork requirements and the bureaucratic process overall – become very apparent. Mpho is a 47-year-old former activist from Mossel Bay who was tortured during political imprisonment:

> I only called on Monday. And the lady told me that they are still busy with my application, and they have been working on it for more than a year already. And on 1 December we applied, last year, so it is almost a year. And I spoke to the lady [...] and after a long wait she came back and she said they are still busy with the verification. And those are the frustrations that we are having. They are still busy with verifying, I mean, we have been involved, we have been in prison, we have been detained, we were … our files were supposed to be quite clear-cut, so what are they still verifying? And now you find … and it also now has a racial connotation to it [...] if you find the guys in the other provinces, I mean they are also like us activists, and they are getting a monthly pension and their families are getting a monthly contribution. Because I have spoken to guys in the Eastern Cape. I mean the old guys, the ex-combatants, they are getting pensions already.

This extract sums up well the huge frustrations felt by people who dedicated their lives to the liberation struggle and keep being told to be patient.

Uncertainty about waiting times and the bureaucratic process can increase feelings of insecurity and powerlessness (cf. Auyero 2012). Like Mpho, a large number of participants reported feeling in the dark about whom to contact about their claim, how long it would take and what exactly the 'verifying' involved. People's experiences of waiting for compensation clearly also involved feelings of competition with other victims/survivors, as is shown in the above reference to ex-combatants getting pensions already. While it goes beyond the scope of this chapter to explore the struggles over recognition and hierarchies of victimhood in post-Apartheid South Africa, unequal waiting times may well contribute to fractures in solidarity among victims/survivors.

Moreover, as the literature on the politics of waiting reviewed earlier demonstrates, waiting can exacerbate social inequalities and affects those that are most marginalised disproportionally more severely (Auyero 2012; Reid 2013). Within this study, this was particularly noticeable in relation to a divide between urban and rural. Rural and urban areas differ in terms of experiences of victimisation and in relation to the recognition and compensation they received after the transition. Martin's story was typical for the respondents I interviewed in the Southern Cape: 'I personally feel very betrayed. Because some of my fellow comrades are now in top positions, and it seems that they don't look back to us. And you must remember that we are in rural areas. Everything is happening in the towns.' The 'looking back' in this excerpt can be read both in a temporal and spatial sense. There was also a strong feeling in other interviews of 'being left behind' (*passim*), a similar spatial and temporal construction. Martin characterises himself as remaining in an essentially unchanging place, while others progress and move forward. To see oneself as 'being left behind' similarly expresses not only a sense of betrayal, but also of anachronism: being made to feel less modern, less advanced or less able to move on.

Feelings of betrayal were more dominant in the rural areas where arguably there has been a slower pace of chance since the end of Apartheid. Participants there felt that they had the longest wait and were most disadvantaged in terms of service delivery, redistribution and patronage networks:

> What also happens was that we are in the rural area. And we are away from town, and most of the guys that were with us they all benefited due to the fact that they are in an urban area. And our problem is that because we are in the rural area, and we are disadvantaged by that. And we felt that if they are in a position to assist us, because they are in a position of power and influence. But they are not using their position of power and influence to assist their comrades now. And that is also the gripe why a lot of the guys left the rural party, because of a sense of betrayal.
>
> (Ariz)

It was also mainly interviewees from rural areas in the south of the Western Cape that reported incidents of racial discrimination, such as the use of racist terminology like 'kaffir' by whites, or being made to feel unwelcome in certain public spaces.

Waiting for change

The experience of waiting was also articulated in relation to change and redistribution. For example, one former political activist in Mossel Bay explained that the lack of socio-economic change brought feelings of bitterness:

> If I stay in a shack, prior to '94, then here comes '94, the country has changed, we are in a democracy, my needs will be answered, and we have been promised that things will change. Housing will be there, electricity will be there, water will be there. But now you are still sitting there, still waiting, for a house, still waiting for whatever. And don't get. And that brings the bitterness.
>
> (Anton)

A victim/survivor from Khayelitsha told me about the long wait for decent housing in this huge informal settlement: 'But even today they are victimising people, even where I stay. We are still waiting, I don't know for how long. But still, we are still like prisoners' (Solomon). Housing is a particularly powerful symbol of the promises and betrayals of the liberation: South Africa's 1996 Constitution, one of the most progressive in the world, enshrined the right to adequate housing. Millions of houses have been built, but not at the pace of population growth (Marais 2011). Just under one in four people continue living without formal housing (National Planning Commission 2012). Overall, the country remains one of the most unequal societies in the world, two-thirds of children live in poverty, unemployment stands at over 35.6% (Statistics SA 2012). Whilst huge progress has been made for some segments of the population, service delivery has been painfully slow for some communities:

> I don't feel happy. It doesn't happen as they put it. As they said, they were going to change things, the only thing they changed is they are doing what white people used to do but now they just take a little bit away from there and add little bit from this one. Really, if you have a clear mind, you can't think it is better now.
>
> (Bongani)

Feelings that not enough had changed or that changes were too slow were very widespread in the interviews, such as Mary, a 63-year-old victim/survivor from Khayelitsha, explains here:

> Nothing changed. Why I feel that nothing changed is because before we were fighting for the Apartheid, we thought when we have a black

> government not a lot of people are going to suffer. But now, as I see it, the people suffer more than with Apartheid. Because a lot of people are not working, there is no jobs for the people. And there are a lot of people in South Africa but less houses, and no jobs, no houses! Only the shacks. The people are living in the shacks. Some people haven't got the taps for water, some people haven't got the toilets. But they keep saying, we are going to create the jobs for the people, and we are going to build the houses for the people. But they are not doing a thing. They just make the promises. For me, the Apartheid years, it was bad but it was a little better than now.

Mary, and many of the other participants in this study, continue to suffer the structural violence they had known under Apartheid, such as lack of formal housing, poverty, lack of access to basic services, and continued de-facto educational and residential segregation (Mueller-Hirth 2017). This can account for why participants often said that 'Apartheid still exists' (*passim*), or, as Gert put it: 'Apartheid is not dead, you see it every day. We still experience it. People think because we suddenly have freedom it's finished. But it's not'. Similarly, the phrase 'I am/ we are still struggling' was very commonplace in interviews to express perceived continuities with the violence of the past. Participants' senses are exacerbated by perceptions about the unevenness of change. In his work, Harms (2013: 348) highlights that 'slowness is relative': residents were waiting for eviction as part of a very slow process of redevelopment of their neighbourhood, while the pace of urban change elsewhere in Ho Chi Minh City was very fast. Similarly, in this study, it is partly the speed of change elsewhere in South Africa that makes any transformation in participants' lives appear too slow or inadequate. This is partly because of the disjuncture between political and economic transformation: the political freedoms of post-Apartheid have not been accompanied by a redistribution of wealth. This gap was a common theme in interviews:

> When you know people have sacrificed with their lives to be liberated and the end result is not what we thought it would be, when the issues that you fought for are not being dealt with, basically. So we have masses of unemployed people, homeless people, large sectors of our population sit at home and wait for welfare pensions and benefits.

The phrase 'sitting at home and waiting' is a significant one: waiting was often coupled with 'sitting' in interviews, highlighting a link between waiting and immobility. In the next section, I will explore how waiting was negotiated. Although immobility appeared to be one of the main effects of waiting, some participants refused to wait or challenged the role of waiter.

The effects of waiting

Sitting and waiting – becoming immobile

As became apparent in the last extract in the previous section, the word 'sitting' has temporal resonances with waiting and is dichotomous to constructions of time associated with change and progress. Munia is a 67-year-old woman and well-known Anti-Apartheid activist in Bonteheuwel, a predominantly coloured township in the Cape Flats. She experienced police brutality during the struggle but also witnessed the murders of two of her grandchildren and one of her daughters due to gang violence in more recent, 'post-conflict', years. She says:

> I am so lonely and I am sitting there alone in the room, and I am thinking back, how it was at that time, and how it is now. And the difference. Because at that time everything was … how to say that, everyone was willing to do something, but now nothing is going.

Waiting is acknowledged to be a significant facet of immobility (Conlon 2011). As discussed in the previous section, victims/survivors shared their senses of there not being enough transformation. Other studies have linked inactivity with victims' difficulties to cope with trauma (e.g. Igreja et al. 2009). By contrast, mobility is associated with movement and change, with 'doing something' as Munia put it. To be sitting and waiting means to put one's life on hold and to be fixed in a state of hope and expectation. Being subjected to waiting renders people unable to make longer-term plans and exert control over the future. This might well prevent them from transitioning out of their victimhood.

In the extreme, prolonged immobility might be seen as giving up hope entirely. S'bu, a pastor in Khayelitsha, told me about some ex-combatants who were struggling to come to terms with the past:

> You can be shocked, these people today they are living just to die, waiting for their day. Given up hopes, because the Apartheid condemns that you must not think as a human being, in other words, frustrates you, everything of yours that may allow you to go further.

Death was invoked in a number of interviews in this manner. Molapo, in Mossel Bay, insisted that 'if I can take you here or there to see for yourself, there are people who are just waiting to die. Not even wanting to do anything. Now they are drinking to die'. Death was also invoked to emphasise the slow pace of change. For example, Martin, who suffered detention and torture and is living in very destitute circumstances, is turning 54 now and has long waited for a house, having fought in the liberation movement: 'My problem is that I am getting older. If I die now, what does my family get? There is not even a house there. My wife, she always tells me, you have put so much in the struggle, you are still putting so much into the community.' It is of course also true that some victims have died before receiving reparations. For example, one respondent who had been forcefully removed from District 6 explained that he had finally been given a beautiful three-bedroom house, but that 'government know a lot of people died already. People are getting older' (Amil). That death might come before victim support has been received serves as a powerful reminder of how much time has passed since Apartheid was overthrown.

Waiting and doing

Waiting is not always linked to inactivity, however. There are multiple temporalities of waiting and waiting can be active (Gray 2011) as well as productive (Auyero 2012). While a sense of immobility in response to delayed reparations or little change was strongest in this data, some respondents challenged the dynamic of waiter/person in power they wait for. An interesting illustration is provided by a group of former activists in Mossel Bay, who spoke about their long wait for meaningful localised symbolic reparations. While huge efforts to memorialise were undertaken in South Africa, by a government that is after all constituted from the former liberation movement, participants felt that this had only commemorated national leaders and did not speak to their local experiences and dynamics. At the time of research, the group were starting a local initiative that sought to acknowledge the activists that were involved in Mossel Bay, rather than continuing to wait for official communication.

Other participants refused to view long delays as the end of their hopes for reparations or transformation. For example, a number of participants discussed the idea of the TRC re-opening its registration process in order to allow victims who had not come forward at the time to be acknowledged:

> So there is also a new move to get the government to reopen the hearings of the TRC again. So that people can come and tell their stories. Because for some of the people, the hearings are not yet completed. Because some people are still having psychological problems pertaining to the struggle, and they haven't told their stories. But we are not too sure if the TRC will be opened.
>
> (Mustafa)

It should be noted that many of the study participants would have met the TRC criteria, but chose not to come forward or were unable to do so. The TRC staff themselves had already identified about 8,000 additional victims who were eligible, but Khulumani indicates that only 10% of its 50,000 members had access to the commission (Interview with Marjory Dobson, 7 March 2011).

A re-opening of the TRC has never been officially mooted. Therefore, victims/survivors' narratives and mobilisations in relation to the commission might be read as a device through which they expressed their refusal to wait and accept that this is the end of their struggle. Waiting is then not passive, but rather an active process in which delays are contested (also see Sutton et al. (2011) who show that migrants waiting for deportation were developing what they call elaborate 'migration stories' to contest the prospect of expulsion). There have also been campaigns for a full implementation of the TRC's recommendations. To mobilise for a re-opening of the TRC can be read as a temporal challenge to the timeline of the official transitional justice mechanism (Mueller-Hirth, 2017).

This is because transitional justice time is linear and teleological, constructing concepts of past present and future that are highly normative. As Alexander Hinton argues, in Chapter 3 of this volume, within the timescape of transitional justice the violent past is delimited and essentialised and set against an imagined liberal democratic future, with the present liminal transitional justice mechanism the means of teleological change. Truth commissions in general are temporally limited (i.e. they last for a fixed amount of time), but this liminal nature of was exacerbated by the TRC's closed-list approach: after a final closing date, applications could no longer be made. Victims'/survivors' narratives and campaigns for a re-opening thus partly challenge this dominant temporality and the power relations inherent in waiting by offering to grapple with the past in a way that cannot be conceived of in teleological transitional justice time.

Fixed Liminality

Waiting can be understood as a liminal condition (Lahad 2012; Sutton, Vigneswaran and Wels 2011) – an ambiguous and uncertain, but *temporary*, in-between stage. Anthropology teaches us that all transitions are marked by three phrases: separation, margin or trial (*limen*) and re-aggregation (van Gennep, cited in Turner 1964). As the middle stage, liminality involves disorientation and a kind of 'structural invisibility' (Turner 1964: 9) of those undergoing the transition. Liminal subjects are 'neither one thing nor another, or may be both; or neither here nor there; or may even be nowhere' (ibid: 7). They are no longer part of the group they previously belonged to, nor are they part yet of the group that they will belong to when they re-enter the social structure. The structural invisibility of those in liminal phases underscores the links of liminality with power inequalities.

The concept of liminality is useful to understanding moments of transition, but it might also be useful in understanding the phenomenon of victimhood within such transitions. Victims and survivors are said to inhabit liminal social spaces; they are 'part of society but removed from society' (Hamber and Wilson 2002: 38). Victimhood itself is usually framed as a temporary and transitory state, at the end of which the survivor emerges as a citizen in the new peaceful and democratic state. Wilson (2001) has, moreover, shown that transitional justice institutions, such as the TRC, are liminal institutions that are supposed to enable the passage from violence time to peace time. They are sites where victims seek to resolve their liminality through legal and truth telling mechanisms. Returning to the above argument that those in liminal phases are structurally relatively invisible and powerless, victims in South Africa have had little leverage to push forward a reparations agenda – and little choice to 'refuse' waiting, partly because they are predominantly poor.

It is hope that enables waiters to endure the liminal state of waiting. As Sutton et al. argue, this is because hope is 'grounded in a longing for and expectation of a new status or identity at the end of the waiting period' (2011: 30). Participants in this research might indeed be considered to be waiting for a new identity in a new democratic South Africa. Their involvement in the freedom struggle and their active role in overcoming Apartheid oppression provided a powerful identity. In the post-Apartheid present, the fact that the new South Africa they had imagined and struggled for has not manifested renders hope so delayed, and increasingly impossible.

The notion of permanent or fixed liminality, however, might better capture the experiences of participants who continue to wait and deal with their victimhood. Given that liminality by its definition describes a temporary state, this might seem paradoxical. However, in contrast to the earlier anthropological literature on liminality, Bauman (1992) has applied the concept of liminality to a post-conflict context – that of Eastern Europe. Liminality might be the dominant phase for a long time, and different parts of society might remain in it for a long time. This is important since, in contrast to Turner's conception of the middle stage as teleological, Bauman highlights the ambivalence of liminality: 'it is a condition without clear time-span, obvious exit and authoritative guides' (1992: 116). Szakolczai (2014) also writes on permanent liminality, claiming that any of the stages of rites of passages could become 'frozen'. Regarding the middle stage I am interested in here, 'it might conceivably happen that a temporary situation becomes extended, lasting, eventually all but a permanent state. This is the familiar experience of "time stood still": an illness that was supposed to last for a few days becoming acute; a war that was supposed to last for a few months dragging on for years and years' (Szakolczai 2014: 34).

What distinguishes fixed or permanent liminality from other liminal phases is 'a fundamental vagueness as to its end point' as Lahad (2012) argues in relation to prolonged singlehood. This insight can be applied to the experiences of victims/survivors in this study that characterised themselves as still waiting, be it for reparations, the special pension or transformative change. Feelings of uncertainty were commonplace, as I documented throughout this chapter. As attempts to ritually create symbolic closure, reparations are 'one key strategy survivors pursue in order to address their overwhelming feelings of uncertainty' (Hamber and Wilson 2002: 44). Prolonged waiting for reparations, redistribution or recognition, in the case of our study participants for two decades, should then be understood as an expression of fixed liminality. Victims/ survivors can become fixed in waiting. Reparations or recognition would allow the resolution of that transitional state and provide closure – but uncertainty about if and when this will be received leaves some victims unable to transition out of their victimhood. The fact that the government continues to delay without entirely destroying hope further consolidates this fixed liminality.

Conclusions

This chapter has shown that waiting – be it for recognition, for compensation or for change – characterises many victims'/survivors' experiences 20 years after the end of Apartheid. It has overwhelmingly negative effects, although a number of participants actively rejected the role of the waiter. I have sought to challenge the notion of victimhood as a liminal and temporary phase by discussing the experiences of some victims in this study as fixed liminality: one of the most significant negative effects of waiting is that some participants are fixed in waiting for reparations or recognition to provide closure. This clearly makes the timing of reparations a crucial issue in post-conflict societies. Moreover, in relation to the aims of this collection, I have shown that theoretical engagement with the sociology of time and the scholarship on waiting can help us to better understand the experiences of victims/ survivors a long time after the formal end of a peace process. When we understand delays and waiting not as neutral but as forms of power and betrayal, we become alert to socio-temporal domination and to the 'double' marginalisation that victims experience.

But why do the powerless continue to wait? There are at least two (complementary) answers to this question: because they cannot afford not to, and because they continue to hope. These reasons, necessity and hope, arguably are not unique to the South Africa case. However, in relation to hope, an additional factor in accounting for the patience and continued capacity for hope among some South African victims might be the bond and loyalty they have to the ANC. The power of this attachment and the memories of the struggle need to be accounted for when trying to understand the dynamics of hope, expectation and compliance. Given that the liberation movement is now the ruling party, some victims/survivors continued to have expectations of the state to provide the support, recognition and change they need. The state continues to 'adjourn without totally disappointing', and so they continue to wait. There is a paradox here: in order to be able to move on from the past, might people have to give up hope of compensation and stop taking the government to task on its promises to victims?

As I was writing this chapter, I was frequently reminded of the drawing of the anxious people sitting and waiting on the sofa in the Waiting Place. In Dr Seuss' story, waiting is not for them and they move on to a brighter place. In the real world of this study, most of the victims/survivors were poor and marginalised, which meant they were made to wait in the first place, could not afford to stop waiting, and the uncertainties of the waiting experience gave them less control over their lives and the future. The waiting place really is a most useless place, but this chapter has sought to highlight that it is also a place in which the powerless must learn to patiently wait.

Notes

1 The support of the Leverhulme Trust in funding the research on which this chapter is based is gratefully acknowledged (grant number F/00 152/AK).
2 Participants were given first-name synonyms. Ages, where indicated, were correct at the time of interview.
3 The amnesty provision was unique to the South African TRC in that it offered perpetrators amnesty in exchange for the disclosure of knowledge of human rights violations. To qualify, perpetrators had to demonstrate they committed these acts with a political objective, made a full confession and show that they received orders from a political party or the state. Once applicants were granted amnesty they could not be charged for the same act criminally or civilly.

References

Adam, Barbara. 2004. *Time*. Cambridge: Polity.

Auyero, Javier. 2012. *Patients of the State: The Politics of Waiting in Argentina*. Durham, NC: Duke University Press.

Bauman, Zygmunt. 1992. The Polish predicament: A model in search of class interests. *Telos* 1992(92): 113–130.

Bourdieu, Pierre. 2000. *Pascalian Meditations*. Stanford, CA: Stanford University Press.

Bryson, Valerie. 2007. *Gender and the Politics of Time: Feminist Theory and Contemporary Debates*. Bristol: Policy Press.

Chapman, Audrey and Hugo Van der Merwe. 2008. Assessing the South African Transitional Justice Model. In Hugo Van der Merwe and Audrey Chapman (eds), *Truth and Reconciliation in South Africa: Did the TRC Deliver?* Philadelphia: University of Pennsylvania Press, 1–21.

Charmaz, Kathy. 1991. *Good Days, Bad Days: The Self in Chronic Illness and Time*. New Brunswick, NJ: Rutgers University Press.

Conlon, Deirdre. 2011. Waiting: Feminist perspectives on the spacings/timings of migrant (im)mobility. *Gender, Place & Culture* 18(3): 353–360.

Department of Justice and Constitutional Development. 2016. President's Fund Report 2015/16 Financial Year. Pretoria: The DOJ & CD, Republic of South Africa. Retrieved 1 May 2017 www.justice.gov.za/reportfiles/other/presfund-anr-2015-16.pdf.

Dr Seuss. 1990. *Oh the Places You'll Go*. London: HarperCollins.

Faull, Lionel. 2011. Special Pensions Unit Fails Activists. *Mail&Guardian*, 30 September. Retrieved 20 April 2017 https://mg.co.za/article/2011-09-30-special-pensions-unit-fails-activists.

Gontsana, Mary-Anne. 2013. Over R1 Billion in Fund – yet Apartheid Victims still Await Compensation. Retrieved 20 August 2015 http://groundup.org.za/content/over-r1-billion-fund-yet-apartheid-victims-still-await-compensation.

Gray, Breda. 2011. Becoming non-migrant: Lives worth waiting for. *Gender, Place & Culture* 18(3):417–432.

Halberstam, Judith. 2005. *In a Queer Time and Place: Transgender Bodies, Subcultural Lives*. London: NYU Press.

Hamber, Brandon and Richard Wilson. 2002. Symbolic closure through memory, reparation and revenge in post-conflict societies. *Journal of Human Rights* 1(1): 35–53.

Harms, Erik. 2013. Eviction time in the new Saigon: Temporalities of displacement in the rubble of development. *Cultural Anthropology* 28(2): 344–368.

Igreja, Victor. 2012. Multiple temporalities in indigenous justice and healing practices in Mozambique. *International Journal of Transitional Justice* 6(3): 404–422.

Igreja, Victor, Wim Kleijn, Beatrice Dias-Lambranca, Douglas A. Hershey, Clara Calero and Annemiek Richters. 2009. Agricultural cycle and the prevalence of posttraumatic stress disorder: A longitudinal community study in postwar Mozambique. *Journal of Traumatic Stress* 22(3): 172–179.

Khulumani Support Group. 2017. TRC Beneficiaries, Victims and Survivors Engage the Staff of the TRC Unit. Retrieved 20 April 2017 www.khulumani.net/reparations/item/1279-trc-beneficiaries-victims-and-survivors-engage-the-staff-of-the-trc-unit-in-the-department-of-justice-momentum-building-323-pretorius-street-pretoria-today-8-march-2017-from-10-00-to-demand-a-fair-inclusive-and-just-reparations.html.

Lahad, Kinneret. 2012. Singlehood, waiting, and the sociology of time. *Sociological Forum* 27(1): 163–186.

Langer, Arnim, and Graham K. Brown. 2016. *Building Sustainable Peace: Timing and Sequencing of Post-Conflict Reconstruction and Peacebuilding*. Oxford: Oxford University Press.

Marais, Hein. 2011. *South Africa Pushed to the Limit: The Political Economy of Change*. London: Zed Books.

Mueller-Hirth, Natascha. 2017. Temporalities of victimhood: Time in the study of postconflict societies. *Sociological Forum* 32(1):186–206.

National Planning Commission. 2012. *Our future – make it work. National Development Plan 2030*. Pretoria: The Presidency.

Parliamentary Monitoring Group. 2012. Special Pensions: Progress report by Deputy Minister of Finance. Retrieved 1 May 2017 https://pmg.org.za/committee-meeting/15253/.

Reid, Megan. 2013. Social policy, 'deservingness', and sociotemporal marginalization: Katrina survivors and FEMA. *Sociological Forum* 28(4): 742–763.

Schwartz, Barry. 1974. Waiting, exchange, and power: The distribution of time in social systems. *American Journal of Sociology* 79(4): 841–870.

South African Coalition for Transnational Justice. 2012. Report: National Dialogue on Reparations 4–6 December 2012. Retrieved 13 January 2016 www.khulumani.net/reparations/item/735-conference-report-national-dialogue-on-reparations-4-6-december-2012.html.

Special Pensions Act. 1996. Retrieved 1 May 2017 www.saflii.org/za/legis/num_act/spa1996182/.

Statistics SouthAfrica. 2012. Census 2011. Pretoria: Statistics South Africa.

Sutton, Rebecca, Darshan Vigneswaran and Harry Wels. 2011. Waiting in liminal space: Migrants' queuing for Home Affairs in South Africa. *Anthropology Southern Africa* 34(1&2): 30–37.

Szakolczai, Arpad. 2014. Living permanent liminality: The recent transition experience in Ireland. *Irish Journal of Sociology* 22(1): 28–50.

TRC. 1998. Truth and Reconciliation Commission of South Africa Final Report. Cape Town: TRC. Retrieved 20 August 2015 www.justice.gov.za/trc/report/.

Turner, Victor. 1964. Betwixt and between: The liminal period in rites of passage. *Proceedings of the American Ethnological Society*: 4–20. Retrieved 20 April 2017 www2.fiu.edu/~ereserve/010010095-1.pdf.

Wilson, Richard. 2001. *The Politics of Truth and Reconciliation in South Africa: Legitimizing the Post-Apartheid State*. Cambridge: Cambridge University Press.

Zerubavel, Eviatar. 1981. *Hidden Rhythms: Schedules and Calendars in Social Life*. Chicago, IL: University of Chicago Press.

8 Time to hear the other side

Transitional temporalities and transgenerational narratives in post-genocide Rwanda

Richard Benda

Introduction

'We felt that it was time to shift the focus from the exclusive stories and listen to the other side.' This statement was offered by a young survivor of the genocide against Rwandan Tutsi to my question about the rationale behind 'Youth Connekt Dialogue' (YCD or 'the dialogues' in the text). YCD refers to a series of public events at district level where children of perpetrators (CPs), aged between 18 and 35 met with children [of] survivors to share their stories of living and growing up in post-genocide Rwanda as emerging and second generations. This contribution comes out of fieldwork research on the first phase of YCD (10 May – 30 June 2013) consisting of fifteen dialogues in the districts of Ngororero, Burera, Rubavu, Nyabihu, Gisagara, Nyamagabe, Nyaruguru, Kamonyi, Bugesera, Kicukiro, Gasabo, Nyarugenge, Kayonza, Gatsibo and Kirehe, as well as Iwawa Youth Rehabilitation centre. The event's slogan was 'Youth Connekt Dialogue: Promise of a post-genocide generation.'

To fully appreciate the significance of YCD, one must place it within the wider context of post-genocide transitional politics and narratives of national reconstruction. A theoretical and normative framework for understanding and evaluating Rwanda's post-genocide recovery has been an ongoing contentious issue among scholars of the transitional period (Straus and Waldorf 2011). However, a reading of *Re-imagining Rwanda* (Pottier 2002), *Remaking Rwanda* (Straus and Waldorf 2011), *Rwanda under RPF* (*Journal of Eastern African Studies* 8(2), 2014), *Courts in Conflicts* (Palmer 2015) and *Making Ubumwe* (Purdeková 2015) provides a comprehensive and balanced starting point for this chapter. More concretely, it is useful to analyse YCD in light of how state-sponsored public events are used to disseminate the post-genocide regime's narrative on national reconstruction (Thomson 2011). Of further relevance is the role of children and young people in this narrative (Pells 2011; Pells, Pontalti and Williams 2014), with a particular emphasis on the agency of different generations of CPs (Mclean Hilker 2011).

This contribution adds to this growing scholarship by attending to a third aspect of this analysis; looking particularly at how the timing of YCD fits within the social and political organisation of time in post-genocide politics. Working from the theoretical background of dominant, privileged and marginalised temporalisations in transitional societies (Valverde 2014; Crawford 2015), it approaches YCD as microcosm of Rwanda's transitional society where plural, competing and complementary temporalities interact constantly and fluidly.

Since the last decade of the 20th century, Rwanda evokes more than just geography or spatial reference. Rwanda has become a story with a timeline dominated by one moment of such magnitude that it has almost become the summary of the whole story. Most accounts of the genocide against Rwandan Tutsi in 1994 tend to start with a historical background or at least a timeline of the events. By linking time and causality, this approach implies that in time and political history, one might find the clues to a better understanding of one of the most brutal and tragic events of modern human history.

Time and the genocide against the Tutsi are intractably linked. No other genocide is locked in a perpetual embrace with a moment in time as the Rwandan catastrophe is with the year 1994. As much as the 'year 94' will be permanently marred by the events in Rwanda, the genocide against the Tutsi would seem conceptually diminished without its specific timestamp. Moreover, temporal references such as 'April 1994', 'April–July 1994' or '100 days' have become customary narrative features in popular and academic accounts.

However, the importance of time as a factor came into sharper focus in the critical transitional period that followed the genocide. This period has been unquestionably dominated by the Rwandan Patriotic Front (RPF), the rebel movement that won the civil war of 1990–1994 and put an end to the genocide. This study interrogates the view held by sections of the scholarship on transitional Rwanda that RPF-led governments detain the monopoly of custody and mastery of time in their project of social engineering (Straus and Waldorf 2011). In her remarkable research on post-genocide reconciliation, Thomson (2013) suggested that at the height of its social and political organisation of time, the state apparatus regimented the weekly lives of Rwandan citizens. Whilst making concession for the state's dominant temporalising, I argue that the organisation of YCD was a critical juncture in the transitional timeline that showed how survivors and CPs bring their own senses of time to bear on transitional temporalisation.

The study also frames YCD as a restorative encounter *sui generis*. In doing so, it explores the timeliness of the emergence and invitation to dialogue of CPs as the politically responsible 'others' who are preferable to unrepentant or untrustworthy Hutu *genocidaires*. CPs' active introduction in the arena of transitional politics signalled more complex forms and experiences of temporality (Thompson 1967). For instance, their shared status with children (of) survivors as 'future oriented-people' (Améry 1980) translated into intergenerational shared time-sense expressed through the metaphors of 'rear-view mirror' versus 'windshield' temporalities. These metaphors will be developed later in the chapter when discussing the directional dichotomy characteristic of transgenerational temporalities in post-genocide Rwanda. At this stage, it is sufficient to point out that they imply dialectic and different trajectories in time-sense between participants in YCD and their adult relatives. Conversely, they broadly align with the government's future-looking temporalisation as expressed by 'Vision 2020' and its focus on children and youth as the 'Rwanda of tomorrow'.

It is argued that the study of YCD through the lenses of time provides a better understanding of how plural, competing and complementary temporalities interact in the transitional timeline, to reorder the past and shape the future. YCD envisioned this future as a 'tomorrow of peace without machetes'; a time when being Rwandese – *Ndi Umunyarwanda* – roots out ethnic identities and their history of misery. This time-focused approach makes it possible to view transitional processes as not happening in a temporal vacuum. Rather, time affects and informs processes and actors; it is a key factor and an invaluable resource in transitional policies. In the specific case of post-genocide Rwanda and YCD, this approach demonstrates better than most how generational narratives and political agencies evolve through and are shaped by time; leading to complex and plural forms of temporality in transitional reconstruction politics.

Some important terms and concepts used in this study need clarification.

- *CPs or 'Children of perpetrators'* refers to the young Hutu of the emerging and second generations.
- *Second and emerging generations*: In this contribution, 'emerging generation' refers to young Rwandans who were aged between birth (including children born during the genocide) and 12 years in April 1994. 'Second generation' comprises young Rwandans born after the official end of the Rwandan genocide.
- *Children (of) survivors*: refers to young Tutsi of the emerging and second generations who survived genocide as children or were born of Tutsi survivors. As much as semantic coherence allows it, the term 'survivor' (active) is preferred to 'victim' (passive).
- *Genocide*: Wherever possible, the events 1994 are given their legal name 'genocide against the Tutsi'. Whenever 'genocide' is used alone; it effectively operates as a shortening of the full name rather than a denial of its specificity.

The content of this chapter is presented in three main sections. The first section documents how post-genocide priorities of security, justice and state building contributed to the rise of a form of dominant official temporality that is future-oriented in keeping with dramatic demographic changes. The second section analyses YCD as an instance and site of subaltern temporality; one that prefers a 'windshield' perception of transitional time to a problematic 'rear-view mirror' time-sense. The final section makes the case for understanding post-genocide Rwanda as a multi-layered transitional society where plural and competing temporalities shape the political landscape.

Transitional priorities and state dominant temporality

The new post-genocide government of national unity sworn in on 19 July 1994 inherited a ruined country and a shattered people. Faced with enormous internal challenges and mounting pressures from an international community slowly emerging from its dismayed slumber, the new government set about the task of rebuilding post-genocide Rwanda with commendable efficiency. This titanic undertaking has received intense and increasingly specialised scrutiny from the research community. There is currently a growing and critically informative literature on transitional policies ranging from territorial reorganisation to land legislation (Newbury 2011; Ansoms 2009; Chemouni, 2014; Ingelaere, 2011), from the prosecution of genocide through conventional institutions – ICTR and national ordinary courts- to the prevention of genocide ideology (Clark 2010; Clark and Kaufman 2009; Peskin 2008; Palmer 2015; Straus and Waldorf 2011), from reconciliation and restorative justice (Thomson 2011; 2014) to vast programmes of re-education on identity and citizenship (Purdeková 2015).

The scale and methods of this post-genocide reconstruction undertaking has led some scholars to speak of a nationwide and ambitious project of social engineering; whether it is seen as a remaking (Straus and Waldorf 2011) or a re-imagining (Pottier 2002) of Rwanda. However, what has received negligible interest in this literature is the actual socio-political organisation of time by successive RPF-led governments as the dominant institution. In situations like post-genocide Rwanda with the peculiarities and burdens of a dualist transitional society (Drumbl 2000), time becomes a vital instrument of political power and significant power tends to rest with 'dominant groups who are able to impose their construction of time on others' (Crawford 2015: 7).

In this context, one can begin to understand the resolve of RPF-led governments to shape Rwanda's post-genocide transitional society and to impose themselves as the sole custodian of transitional temporality. At least three main reasons are behind this quest for temporal dominance: security, the prosecution of genocide and the need to build a strong post-genocide state. The latter explains another characteristic of governmental temporalising, namely its overt focus on the future as documented in 'Vision 2020'. Dominant in nature and predominantly future-oriented in direction, the transitional governments have sought to impose this organisation of time not only nationally but vis-à-vis the international community as well.

Let us consider the prosecution of genocide as an example. Given the nature of this crime, it was inevitable that both national and international courts would claim priority of jurisdiction. On 8 November 1994, the United Nations Security Council (UNSC) passed Resolution 955 which created the International Criminal Tribunal for Rwanda (ICTR) based in Arusha. Palmer has noted that when Resolution 955 was adopted, Rwanda was the only but important dissenting vote (Palmer 2015: 45). Therefore, it came as no surprise when, in response to the UNSC Resolution, the Rwandan authorities passed

Organic Law 8/96 which instituted 'Specialized Chambers' within national courts to prosecute genocide-related crimes.

Right from the start, the two judicial institutions disagreed on a very fundamental point, namely their jurisdiction *ratione temporis.* More concretely, the UN and the new Rwandan government diverged with respect to the period over which crimes considered as constituting genocide had been committed. The ICTR's mandate would only cover genocide-related crimes committed between 1 January and 31 December 1994. However, Rwandan authorities argued that temporal jurisdiction should cover the period between 1 October 1990 (the start of the civil war) and 31 December 1994. This duality has remained unresolved throughout the delivery of transitional justice in post-genocide Rwanda (Webster 2011).

The significance of this disagreement over temporal jurisdiction should not be understated. It is further evidence of the determination of Rwandan authorities not to be at the mercy of external forces; especially an international community that had stood by whilst the Tutsi were being systematically annihilated (Melvern 2000; Dallaire 2003). Furthermore, by asserting their jurisdiction over the temporality of genocide, Rwandan authorities sought to underline their status as 'victors'. They would not accept to be cast as losers or victims who would relinquish the control of post-genocide transition and narrative in the hands of some amorphous international community.

Thus, the BBC found this out at their expense when their *Rwanda: The Untold Story* documentary (2014) was met with withering criticism by the Rwandan government, culminating in the suspension of all BBC broadcasts on Rwandan territory. I have argued elsewhere (Benda 2014) that the Rwandan authorities perceived the audacity of the BBC to put forth an alternative story as an infringement on the boundaries of 'narrative' competence. It was also a challenge to the authority of the Rwandan state and people in matters regarding their history. In other words, even if there were any truth in some aspects of the untold story, it is up to the stewards of the overall transitional narrative to determine the substance and the timeliness of its most salient components.

Vision 2020: future-oriented temporality and transitional demographics

The three pillars of governmental temporalisation (security, justice and nation-building) have at times forced transitional governments into a tricky balancing act between past reality and future concerns. Crawford has asked the pertinent question of whether transitional justice can look simultaneously to the past and the future, whilst accommodating both justice and the demands for future security (2015: 10). In post-genocide Rwanda, there have been palpable tensions between RPF-led governments' future-driven temporality and the survivors' need for confronting of Rwanda's recent past. This tension is reminiscent of Shearing's concept of 'two moralities' that inform the governance of post-conflict security and justice. One morality functions from a past-focused logic that seeks to undo the damage done, not only by making material reparations but also by re-establishing the 'mystical balance' of a symbolic, societal order that has been violated. The second morality operates from a future-oriented logic that is less interested in futile attempts to reorder the past and insists instead preserving social cohesion by reducing the likelihood of future recurrence (Shearing 2001: 208–209).

Despite inevitable compromises, a punitive *Gacaca* among others, it is unquestionable that the thrust of official temporality has been explicitly – albeit not exclusively – future-oriented. For the most part, the past has been deployed in the service of, and subsumed under the priorities of security and nation-building (Crawford 2015: 10–11). For their part, survivors have felt that a past-reordering 'justice and what it required was independent of, and a precondition for, the governance of security' (Shearing 2001: 209).

It is important to note that Rwanda is not a unique case in this respect. In his wonderful essay on '*Resentment*', Améry offered a profound analysis of time and victimhood. He noted that contrary to the victims whose time-sense is fixated to the past, the rest of society tends to side 'with those to whom the future belongs'. Future, he goes on to say, is a value concept; 'what will be tomorrow is more valuable than what was yesterday. That is how the natural feeling for time will have it' (Améry 1980: 76; also see Rosoux, Chapter 2 in this volume, on Améry and the right to resentment).

With this in mind, the transitional regime's *Vision 2020* is essential to understanding time organisation in post-genocide Rwanda. It should not be seen merely as an economic framework to transform Rwanda into a middle-income nation (Republic of Rwanda 2012). Instead, it represents the clearest articulation of a highly institutionalised transitional temporality. It indicates the determination of the Rwandan state to lay down temporal landmarks by which it will assess and evaluate post-genocide reconstruction. According to the same official document, 'Vision 2020 constitutes a bond that holds Rwandans as a people determined to build a better future … Rwandans will be a people sharing the same vision for the future and ready to contribute to social cohesion' (ibid: i.9).

This focus on the future and the people who will own it leads inevitably to a brief examination of the impact of time on post-genocide demography. The most remarkable temporal effect of the transitional period has arguably been the dramatic transformation of the Rwandan population. Despite the sketchiness of census data immediately before and after genocide, it is now a matter of consensus that the last two decades have seen the Rwandan population more than double, standing at approximately 11 million in 2013 up from 5,57 million in 1995 (UN Secretariat 2016). Democratic liberalism or lack thereof notwithstanding, it is clear that the different policies implemented by successive transitional governments, especially in the security and public health sectors, have played a key role in this demographic flourishing.

More than the increase in the population, it is the makeup of transitional demography that interests this work. As of 2012, approximately 70% of the

Rwandan population were young people of 30 years of age or under (Pells, Pontalti and Williams 2014). More importantly, at least 50% of this young population were born after the genocide of 1994 (Permanent Secretary MYICT 2013). A cursory look at these figures would suggest that any successful policy must consider this population carefully, given their relevance for the present and the future of Rwanda.

Burman (2008) noted that periods of uncertainty and transition are marked by an increased tendency to symbolically position children in nation-building narratives. In her study of post-war Uganda, Cheney (2007) noted intriguing conceptual links between childhood and nationhood. She suggests that children's connection to the nation both strengthens their position as agents of social change and points to their subservience to adult ideals about both childhood and nationhood. In this context, children are cast in metaphors that romanticise them such as 'pillars of the nation', 'children as the future', while they still embody an idyllic past in the minds of adults (Cheney, 2008: 10). Cheney's work has inspired scholars researching the situation and agency of children in Rwanda's post-genocide reconstruction narrative. In her research on the National Summit for Children, Pells sees Rwanda as a paradigmatic case 'where language and symbolism around children are central to a new narrative of national rebirth' (2011: 78). It is at this intersection between transitional childhood (or youth) and nationhood remaking that YCD has to be located and analysed.

YCD: transitional identities and transgenerational temporalities

Organisers described the YCD's aim using the interesting directional metaphor of transitioning from a 'rear-view mirror' to a 'windshield' perspective (MM, Gatsibo, June 2013). They explained this metaphor as a shift from survivors-centred accounts to alternative stories (TE, Kicukiro, May 2013). In no way does this shift imply narrative replacement or supplanting. Instead it brings into focus and invites worthy interlocutors from the privacy of their feelings into the public arena of transitional politics. YCD constituted one of those key moments in transitional periods when one group acknowledges the other as essential for the (hi)story of the conflict that led to that moment to unfold fully. Therefore, the dialogues offered themselves to interpretation first as a restorative encounter of a particular kind.

Timeline and timeliness of a (trans)generational restorative encounter

The concrete emerging and timeliness of the 'moment of dialogue' (Collins 2004) resulted from a process that has a timeline of its own. As one might expect, the starting point was April 1994 when BE, a hospitalised 9-year-old Hutu boy witnessed the brutal murder of fellow patients because they were Tutsi. After his release from hospital, he was dismayed to find that most of his Tutsi teachers and schoolmates had been killed. Faced with a wall of silence from Hutu adults in his attempt to understand this still localised tragedy, BE used poetry, music and acting to deal with his sense of loss and grief.

Between 2004 and 2006, his artistic work came to the attention of people in government and survivors' associations because of its content – sympathy for Tutsi victims and overt indictment of Hutu adults for genocide. Often invited by survivors to recite poetry as part of their mourning and commemoration gatherings, he was also given the opportunity to share his own traumatic experience of witnessing genocide as a child and growing up with the guilt and shame of a 'child of killers'. As a bond of mutual understanding developed between him and survivors, BE felt that adult Hutu perpetrators should follow his example. Unfortunately, he found that 'for them genocide was like a football game that had come to an end … life should continue and people should move on' (BBC 2013). However, some CPs as well as young Tutsi survivors, especially fellow artists, responded positively to his message.

> We came to the conclusion that we should come together to form the pact of a generation, a pact for life. This is how 'Art for Peace' was born; to create for ourselves a future of peace without the machetes.
>
> (BE, Kicukiro, May 2013)

During the 'Youth Week' in early 2013, Art for Peace invited the First Lady and shared their vision with her. She offered to support their effort to reach out to young people in every district of Rwanda in collaboration with the Ministry of Youth and ICT (MYICT). Thus was born the YCD initiative (Bamporiki 2010: 12–16).

From this brief outline of its inception and timeline, YCD appears as an atypical restorative encounter. A typical restorative encounter brings together the offender and the offended in a justice-oriented moral space for confronting the past in order to govern the future (Shearing 2001). YCD was atypical because it was in actual fact an encounter between two groups who were victims of the *genocidaires*' crimes, although they have been affected differently.

Whilst the victimhood of Tutsi is somewhat easy enough to establish, that of the CPs is more complex and requires careful examination of the psychology of transgenerational transmission of guilt, blame and liability in the particular context of post-genocide Rwanda. Stories told during the dialogues revealed that some CPs are victims of the traumatic witnessing of their adult relatives committing incomprehensible atrocities. Those who were too young or unborn to be witnesses suffer from the silence of their adult relatives as they try to understand the social stigma attached to their lives. In post-genocide Rwanda, the terms 'children of Hutu', 'children of perpetrators' or 'children of killers' are routinely used interchangeably (YCD-Kicukiro, 2013). In a culture where genealogical affiliation and generational ties play a crucial role in social interactions, there is a natural tendency for Tutsi survivors to assimilate children to their progenitors through a blanket transfer of transgenerational or collective blame.

In this context, YCD has to be seen as a timely restorative justice encounter in which these two groups sought to attend to justice in at least two ways. First, both groups were seeking to do justice to each other through shared stories of victimhood. Secondly, there was a clear intention to address the 'transgenerational injustice' suffered by CPs by making it clear that the criminal liability of parents and adult relatives does not extend to their descendants or younger relatives. In the future-oriented context of post-genocide transitional politics, YCD was primarily a platform for the restoration of CPs to their rightful place as blameless Rwandan citizens. On the other hand, it made it possible for survivors of genocide to enter into a civic partnership of national reconstruction without having to compromise with what is impossible to reconcile with, at least for the time being.

From rear-view mirror to windshield: trajectory of transgenerational temporalities

The founders of Art for Peace had a vision for their generation: a future of peace without machetes. The twisted irony of the poetics of this vision will not be lost to speakers of the Kinyarwanda language: *amahoro* (peace) rhymes with *imihoro* (machetes – singular form *umuhoro*), by far the tool of choice in the technology of genocide against the Tutsi. This is not just a vision of the future; it is first and foremost a clear indictment of the past. Transitional temporalities from the perspective of emerging and second generations operate from this stark dichotomy that functions like a truncated caesura: *past is imihoro||future is amahoro*. From this perspective, it makes perfect sense for organisers and participants to locate their political agency in the sort of temporality that wants to turn away from past-saturated narratives (rear-view mirror) to focus on new stories of change (windshield). As one of the young organisers rhetorically pointed out, 'if you drive with your eyes glued on the rear-view mirror, how will you avoid accidents and collisions?' (MM, Gatsibo, 2013).

Rear-view mirror: problematising past-focused temporalities

The implications of this rhetorical question lead to the examination of what is perceived as problematic with past-oriented transitional temporalities. After all, the history of Rwanda is replete with 'accidents' and 'collisions' that resulted from ill-thought politicisation of the past. YCD's stated desire to 'move on' from a *past of imihoro* should first be seen as externalising a more traumatised relationship with this 'machetes season' (Hatzfeld 2006). For many participants in the dialogues, *imihoro* still conjure up moments of chilling horror and personal tragedy (Rieder and Elbert 2013). For young Tutsi participants like KE (Gatsibo, May 2013), physical scars spoke for themselves. For young Hutu like BE, it is the vivid memory of a machete-severed head coming to rest at the foot of his hospital bed. Overall, the reoccurrence of *ihahamuka* (Taylor 2015) episodes during YCD meetings showed that young Rwandans are still gripped by the psychological effects of this past.

Several studies have highlighted the lasting effects of genocide and war (Rakita 2003; Reider and Elbert 2013) as well as the prevalence of PTSD and other psychological disorders among Rwandan young people (Sydor and Philippot 1996; Dyregrov et al. 2000; Munyandamutsa et al 2012; Neugbauer et al. 2009; Schaal et al. 2011). As early as 1995, UNICEF reported that 'almost all Rwandan children have either been victims or witnesses of massacres or they have lost a family member during the 1994 events' (Sydor and Philippot 1996: 235). Sydor and Philippot's research concluded that among other things a 'humanitarian catastrophe can generate psychological reaction even among people who were only witnesses or onlookers' (1996: 242). With this scientific evidence in mind, turning away from 'rear-view mirrors' can be interpreted as an intentional and resilient act of reclaiming one's sanity, health and life, no less. It is daring to imagine lives that are not utterly dominated by the tragedy of 1994.

This desire to reclaim one's life raises fascinating transgenerational paradoxes in both groups. Starting with young Tutsi participants, it was obvious that the majority had been alive at the time of genocide; thus they are as much survivors as their parents or adult relatives. One of the reasons put forward by the organisers of YCD was that post-genocide transitional narratives and rehabilitative policies had been overwhelmingly survivors-centred. In a country of perennially limited resources, being at the centre of political action meant access to at least some of these meagre resources. It also meant keeping the genocide against the Tutsi firmly anchored at the centre of public life. Social bodies and institutions such as the Fond d'Assistance aux Rescapés du Génocide (FARG), the Week of National Mourning/*Icyunamo*, Memorial sites/*Urwibutso* and the different laws against genocide and genocide ideology embodied this survivors-centred transitional temporality.

Why then would young Tutsi survivors and children of survivors contemplate an alternative temporalisation? One might recall Crawford's suggestion that young people's shorter time horizons imply different temporalities (2014: 20). This would imply that young Tutsi have the ability to 'move on' quicker than older generation survivors. Although this might be true, there was little evidence of this assertion during YCD. Instead, one might suggest that these contrasting temporalities hint at a gradual and rather pragmatic re-evaluation of the socio-political dividends of Shearing's 'two moralities' alluded to previously – the benefits of reordering the past on one hand and planning for a cohesive future on the other.

There was a perceptible feeling among YCD Tutsi organisers that past-focused and exclusively survivors-centred temporalising might lead to increased or generalised resentment within the population. Améry (1980) has expertly written about transitional resentment, especially from the perspective of frustrated victims. Resentment, Améry suggested, is past-obsessed. It leads to a twisted and dis-ordered time-sense. It blocks the exits to a genuine human dimension, namely the future (ibid: p. 69). Therefore, it is unappealing to young people who are future-oriented, 'those to whom the future belongs' (p. 76). YCD's repudiation of 'rear-view mirror' temporality implied that the cyclical return to the past might have more value for the older generations of both ethnic groups. Young people should cast in their lot with fellow future-oriented people.

The thought of reclaiming lives from resentment-inducing temporalities was more urgent among CPs. For this group, transgenerational transmission of guilt – moral, societal and even legal – was voiced as a recurrent daily life experience. As one participant put it,

> Being a Hutu child is being a child of killers. You know it and you know that the survivors cannot help but see you in that way, even if they never say it to your face. Who could blame them? For us it is a problem. It is bitterly frustrating but it is what it is.
>
> (UC, Kicukiro, May 2013)

Therefore, by focusing on crimes committed against Tutsi in 1994, 'rear-view mirror' temporalities also focus on their parents' and adult relatives' crimes, and by relational-genealogical association, to CPs. The move to legally ban ethnic divisionism (Law No 47/2001 of December 2001) has done little in reality to minimise transmissibility and collectivisation of guilt across Hutu generations (Rubavu, May 2013). From a *bona fide* wronged party, these young people have become putative perpetrators. This, YCD organisers feared, is the disfiguring effect of rear-view mirror temporalisation in post-genocide Rwanda; a distortion that cuts through the phenomenology of identity for the majority of CPs who attended YCD. They stated that being 'lumped together' with Hutu killers and criminals led to their being held back socially and economically (Gisagara and Kamonyi, May 2013). As an example of this obstruction CPs mentioned access to education, which is only free at primary school and rather expensive at secondary and tertiary levels. Whilst young Tutsi survivors can get support from FARG, their Hutu counterparts have no equivalent, even for those who have both parents in prison (Gasabo, May 2013).

In the course of YCD, it became increasingly clear that for the most part CPs' frustration and resentment were directed primarily toward adult Hutu. Bamporiki's *Their sin, our shame* (2010) captures these feelings succinctly. In the course of the dialogues, CPs spoke of years spent trying to extract the truth from their imprisoned parents without success. One young man was so exasperated by his incarcerate father's denial that he used his last visit to sever his relationship with someone he called a 'disappointing coward' (Kamonyi, May 2013). Finally, there were the painful stories from the most authentic children of killers, namely those born from the raping of Tutsi women by killers. One of them described their condition in the matter-of-fact tone of people for whom emotional pain has become a life companion:

> I do not know who my father was but I know what he was. He was one of many killers who raped my mother. Try growing up with that! My mother has wished me dead from the first day. No family wants anything to do with me and the government does not recognise me as a survivor's child. We are children of sorrow, born to afflict everyone.
>
> (NJ, Gisagara, May 2013)

However, frustration and quiet resentment did not go only in one direction. Cross-generational research shows that this is rarely the case (Berger and Berger 2001; Dan Bar-On 1991). Some CPs were concerned that the 'annual rewinding of time back to April 1994' led to cyclically renewed resentment and brooding animosity between Rwandans. They generally understood that institutions such as *icyunamo* and *Urwibutso* fulfilled the dual role of remembering-*kwibuka* (victims and for victims) and reminding-*Kwibutsa* (perpetrators and for perpetrators). However, they felt that survivors – and the government – used the latter as a means of keeping all Hutu on the defensive politically. This, they felt, was counterproductive in the long term; especially when no transgenerational differentiation is made between perpetrators and their children.

Windshield perspective: committing to a future of peace and responsible citizenship

The problematic relationship between emerging and second generations, and past-focused temporalities went a long way to generate the kind of 'time-feel' expressed by organisers of YCD; 'we felt that it was time to hear the other side's story'. Time feeling in post-genocide Rwanda requires not only good social intuition but also skilful reading of the transitional political landscape. In the case of YCD, it required timing in order to gauge the receptiveness and readiness of the survivors. It also involved generational prioritisation; to hear the perpetrators' side without necessarily listening to the perpetrators themselves. Windshield temporalisation thus implied direction and demographics. As Améry would have it, it grants the 'advance in trust' due to future-oriented persons who do not lay claim to innocence vigorously and impudently, but are instead prepared to bear responsibility for the April–July 1994 catastrophe (1980: 76).

Political responsibility was central to the windshield-facing transition advocated by YCD. It was essential in order to move on from reaction, guilt and endless introspection toward constructive action and change. It was not a vacuous ideal either. The dialogues revealed fascinating stories of CPs taking responsibility for their adult relatives' actions. There was the story of UJ, a young Hutu woman whose brother had been sentenced in absentia, who decided to take his place in TIG (work of general interest) for two years (Kirehe, June 2013). There were also the stories of CPs who set up clubs to help survivors who have lost all their children (Gisagara, May 2013, Kirehe,

June 2013). There were people like BE who held deep-seated empathy for survivors' experiences and stories. Arguably the sort of intragenerational networking that led to the creation of Art for Peace and the organisation of YCD is an instance of political responsibility in action.

Therefore, the organisation of YCD marked a point in transitional temporalising when the time felt right to move these instances of political responsibility from the private to the public realm. It was not only a moment for the externalisation and actualisation of a lingering conflict between Tutsi survivors and Hutu perpetrators in order to master an opposite yet shared past (Améry, 1980). Rather, YCD constituted a moment of concretisation for the 'others' and their stories, as well as a starting point for their incorporation in the overall narrative of post-genocide reconstruction as envisioned within the government's Vision 2020. This last point, coupled with what has been said earlier about governmental future-oriented temporalisation, explains why the government was keen to sponsor and incorporate a bottom-up initiative; particularly one that finds its inception within perpetrators' quarters.

Windshield-view, future-looking and change-driven, this is the new stage in transitional temporality heralded by YCD. Change, it has been elucidated, will be brought about through responsibility to own the past and build the future. Concepts such as 'children of killers', 'children of perpetrators' along with the already outlawed 'children of Hutu' will give place to 'builders of a nation'. YCD served up memorable songs but none was as captivating as Masamba's *Jenga, jenga taifa lako* (Build, build your nation) or Gatsibo youth's *Rwanda itajengwa na sisi vijana*. (Rwanda will be built by us the youth). It is important to emphasise that these 'future builders' – note the passive form and future tense of the Kiswahili verb 'kujenga' – are not just building a country. This is where borrowing from Kiswahili is useful; there is a significant narrative difference between *taifa* (nation) and *nchi* (country). Thus, the next stage in the rebuilding of post-genocide Rwanda will only be possible through the intergenerational solidarity of those to whom its future belongs.

Building a new nation in which the responsible other is not only heard but also fully included in national reconstruction requires a new framework which will determine how time is organised in the next stage of transition. YCD provided such a framework in the form of *Ndi Umunyarwanda*, literally 'I am Rwandese'. *Ndi Umunyarwanda* arose from and is seen as a seamless continuation of YCD (Kagame 2013). It was launched officially on July 2013 and on 23 October 2013, the Council of Ministers decreed that *Ndi Umunyarwanda* would be discussed in all public institutions, civil society, NGOs and religious associations. Similarly, Rwandan Ambassadors and High Commissioners have the mandate to promote this programme among various Rwandan communities of the diaspora (Umuseke, October 2013). As recently as January 2017, a prominent member of the Rwandan Parliament referred to *Ndi Umunyarwanda* as the 'story, vision, philosophy and programme for a new nation' (MP Gatabazi, London, 2017). Through the windshield view, *Ndi Umunyarwanda* promises clearer visibility for the road ahead.

YCD's narratives and plural temporalities in post-genocide Rwanda

In his typology of post-genocidal societies, Drumbl makes a robust case for the classification of Rwanda as a dualist post-genocidal society (2000: 1237–1252). However, closer scrutiny reveals post-genocide Rwanda as a rather plural transitional society in which multiple stakeholders vie to assert their narrative and temporality; in other words their political agency. YCD was a microcosm of this transitional society and the various temporalities that have been at the heart of transitional politics in post-genocide Rwanda. 'We felt that it was the right time to hear the stories from the other side' encapsulated these concomitantly competing and complementary temporalities.

It has been shown that the Rwandan state, through various RPF-led governments, and genocide survivors fall into the category of the keepers or custodians of transitional time, although the former exerts more dominance than the latter. The perpetrators and their descendants constitute the 'other side', cast as those who contribute to post-genocide temporalisation from a position of marginality, subversion, expectation or outright passivity.

As suggested earlier, three main reasons have contributed to the state's dominant temporalisation: security, the prosecution of genocide and the need to build a strong post-genocide state. This has led to a complex type of institutional temporality that sought to hold past, present and future in precarious balance. Valverde (2014) has made a case for the crucial role of temporality in all security projects. This is particularly relevant in the aftermath of genocide in Rwanda where internal and external factors had combined to create a situation of clear, immediate and explosive danger. However, whilst security was undoubtedly a priority so was making sure that all *genocidaires* were immediately neutralised, apprehended, incarcerated and tried lawfully.

Equally requiring justice were the returning Tutsi exiled following the 1959–1964 and 1973 pogroms; the same people who had supported the RPF's war effort and therefore entitled to housing and land. This situation imposed urgent demands on public and civil justice and necessitated delicate temporalisation focused on recent and remote political history. To compound the transitional equation, prosecuting genocide and resettling Tutsi returnees confronted the government with the real possibility of doing a great deal of injustice to – and thus alienating – many innocent Hutu. Since the latter would be an essential and integral part for a united post-genocide state, transitional temporalisation had to be future-minded as well. At different stages of transitional politics, priorities have dictated appropriate temporalities.

Therefore, the decidedly future-oriented temporality displayed by the government through the sponsoring of YCD was in no way indicative of one-dimensional temporalising. Instead, it represented a timely response to a transitional priority and demonstrated the kind of flexible temporalisation that has been characteristic of the Rwandan transitional regime.

YCD also demonstrated that despite its dominant role, the state is by no means the only player in the field of transitional temporality. Since the RPF regime is credited with stopping the genocide against the Tutsi and because it has been labelled 'Tutsi' in many quarters, it has become almost natural to think that their time-sense would align with that of Tutsi survivors. It has also become increasingly acceptable in some academic circles to portray the post-genocide RPF-led regime as strongly authoritarian, even totalitarian (Reyntjens 2004; Purdeková 2011, 2015; Longman 2011). It therefore has the ability to impose its own agenda over the transitional period unchecked, including in all genocide-related matters.

However, YCD showed that survivors play a key role in the decisions regarding the timeliness of policy and other public initiatives on genocide and its legacy. As the prime victims of the tragedy, their recovery and readiness to move on to whatever next stage is required by justice, reconciliation and reconstruction programmes is essential. The organisation of YCD showed that survivors had a strong say about inviting and welcoming the other side's stories. In fact, one can go as far as suggesting that there has been a semi-concerted effort between government and survivors to avoid a transitional situation where, in Améry's words, the public consciousness loses the memory of genocide and the disquiet disappears among perpetrators (1980: 71).

By insisting that genocide be prosecuted for as long as it takes for every perpetrator to be brought to justice, survivors refused to be subjugated to natural, biological or social time-sense (Améry 1980). In this kind of transitional temporality, victims are forced by social pressure to forgive and forget as soon as the crime is considered as being long past (1980: 71–72). Instead, throughout the transitional period, they fostered a kind of 'constructive resentment' among the survivors aimed at arousing 'self-mistrust' in the camp of perpetrators with the expectation that spurred by the former's resentment, Hutu perpetrators 'would remain sensitive to the fact that they cannot allow a piece of their … history to be neutralized by history but integrate it' (pp. 77–78). This, Améry argues, is the appropriate transitional time-sense, one that is moral in essence and redeeming in political intentionality (p.79).

Framing post-genocide temporality in moral terms might explain the timing of YCD and why CPs were the more 'acceptable' choice. A question of constant concern for Tutsi survivors is whether a *génocidaire* can be truly reformed and trusted again, at least politically. Stories of genocidaires who confessed and publicly apologised for their crimes are not uncommon, especially after the Gacaca trials (Clark 2012). Quite beside the near impossibility of ascertaining their authenticity, confessions and apologies can only be considered as the first steps on a long journey to moral and civic rehabilitation. Therein lies the question of time and restorative politics. How long would such a process take? Is it fair on survivors to be expected to go the proverbial extra mile and wait for this conversion – essentially contingent- to materialise?

It is important however to emphasise that the form of non-institutional temporality espoused by adult Tutsi survivors is far from monolithic and goes beyond the moral sphere. These survivors are far from 'uncle Yans' and 'aunty Tans' suspended in a specific time (Hinton, Chapter 3, this volume);

i.e. the time of genocide. Reducing survivors to a one-dimensional time-sense would in fact be undercutting their very identity and experience. Surviving understood as 'living beyond' requires resilience and adaptability that makes life in the aftermath not only possible but also worthwhile. For survivors of the Rwandan genocide this meant seeking justice but also loving, having families and raising children, working and setting up businesses; in other words, celebrating life in the present whilst securing a future for themselves and their descendants. The latter is not different from the post-genocide state's determination to build a Rwanda fit for children (Pells 2011). This meaningful living in the aftermath also meant adopting a flexible relation to transitional time as well as engaging in a complex of interactions with the temporalities of other transitional agencies.

Therefore, the past-focused pursuit for justice does not preclude survivors from Shearing's future-looking logic or eject them from Améry's genuine human dimension, the future. For instance, the existence of reconciliation villages in which survivors willingly live alongside perpetrators (Mugabo 2015; Ong 2017; Baddorf 2010), as improbable as their future may be, proves that survivors are willing to extend trust credit. Those for whom the weight of history and experience of betrayal have made wary of hasty conciliatory discourses have not let these reservations encompass the CPs as well. By supporting YCD and subsequently *Ndi Umyarwanda*, survivors have opted for descendant-focused transgenerational restoration, instead of maintaining the ascendant-inherited transmission of blame. In this respect, they come to the future-oriented, windshield-like temporality championed by government-sponsored YCD but on their terms and following an autonomous process.

Speaking of these 'descendants', YCD showed signs that CPs are far from passive beneficiaries of a gratuitous political credit. If anything, their stories revealed an equally complex time-conscience, one that is finely attuned to the fluidity of transitional politics. For instance, the lobbying for and organisation of the dialogues served as indication that this generation is increasingly aware of their political relevance. It was the right time for them to alert the transitional *agora* to their alternative identities and life stories, which are distinct from those of their progenitors.

Furthermore, although their understanding – and for the most part support for – transitional survivor-sensitive temporalisation was reaffirmed in the course of the dialogues, it was not uncritical. For instance, they questioned the selective nature of the 'past' – the officially edited version of history that informed governmental temporalities. Central to their criticism is the question of 'how far back in time should we go to establish causality and responsibility for genocide?' In Nyabihu and Burera districts (May 2013), young people made a case of historical responsibility against Rwandan Tutsi monarchs for their collusion with (German and Belgian) colonisers. In this case, they inquired, does it stand to reason to blame colonial authorities only and exculpate Tutsi kingship?

By raising these historical questions, young people aimed their criticism against a perceived revisionism of history that seems to sanitise political history from the rise of the Nyiginya dynasty to the troubled period of 1959–1964. This process effectively exonerates all Tutsi regimes and lays historical blame on the Hutu leadership of the post-independence years (1962–1994). CPs voiced concerns regarding this dichotomic segmentation of Rwanda's political timeline and called for a more 'balanced history' to be taught in schools and public sessions (Gisagara and Rubavu districts, May 2013).

CPs' stories also indicated that past-focused and victimhood-centred temporalities of the transitional period had greatly segregated victims. Only Tutsi had been considered victims, leaving out other categories of victims such as children from mixed marriages, Hutu widows and widowers, and Hutu victims of the killings in April–July 1994. Equally overlooked were Hutu killed or widowed by RPF soldiers during the civil war (1990–1994), during the Insurgents' War – *intambara y'abacengezi* – in the north-west of Rwanda (1996–1997), and in the course of the dismantling of refugee camps both inside Rwanda and in the former Zaire. Some young people, especially from the northern districts of Burera, Rubavu and Nyabihu, went as far as subverting the timeline of genocide itself by openly and candidly saying that genocide in fact happened in 1996–1997 (Art for Peace, 2013). This point brought the issues of victimhood temporality back at the centre of public debate. More particularly, it raised the question of when crimes committed by the current regime will be addressed and what would be the modalities.

A final example that showed the desire of CPs to stake their claim in transitional narratives and temporalisation was the liberal use of the officially banned ethnic terms 'Hutu' and 'Tutsi'. Lively discussions of ethnicity showed that the topic is far from settled as implied by laws and policies against divisionism and genocide ideology. Participants pointed to the obvious idiosyncratic and anachronistic nature of suppressing ethnicity whilst maintaining it within the nomenclature of genocide. Both in Burera and Nyabihu districts (May 2013), it was argued that to thoroughly follow the logic of ethnic suppression, the genocide against Tutsi should be known as 'the genocide against people who *were* Tutsi by people who *were* Hutu' (my emphasis). In other words, genocide cannot remain in the present whilst the protagonists are relegated in the past. YCD participants felt that a more community-based and frank dialogue on the topic is preferable to peremptory institutional suppression.

Consequently, a shared national identity would emerge gradually through a process of careful identification of core Rwandan values lost or diluted by colonialism, as well as the responsible owning up to the more recent history of ethnic violence. Thus *Ndi Umunyarwanda* developed out of these dialogues as the quest and framework for the future ethnic-free *umunyarwanda* as envisioned by those to whom that future belongs, both biologically and politically.

Conclusion

When young Rwandans of the emerging and second generations came together in May-June 2013 for intragenerational dialogues, theirs was a desire to make a pact for life with the seemingly straightforward terms of leaving a past of *imihoro* behind and embrace a future of *amahoro*. Yet, a time-aware approach to these dialogues reveals unsuspected ambiguity and complexity. From this perspective, YCD offered a narrative rebuttal to normative transitional binaries survivor-perpetrator and past-future. Thus, by introducing ambiguous identities, complex narratives and fluid temporalities, YCD find their home in what Hinton's contribution to this volume (Chapter 3) calls the 'gray zones' of the transitional period.

Paying attention to the significance of time in transitional processes such as YCD in post-genocide Rwanda makes it possible to prove that there rarely is a single master or custodian of the transitional time. Instead this period is shown to be the site of plural, contested, competing and complementary temporalities as the Rwandan state, genocide survivors and young adults of the emerging and second generations vie to assert their political agency. Alongside the diversity of transitional agencies, directionality equally contributes to plural temporalities. No one transitional stakeholder in post-genocide Rwanda is confined to a specific time. Instead they all entertain a vari-directional and flexible relationship with transitional time; moving fluidly between future-oriented and past-focused forms of temporality in response to transitional priorities or necessities.

YCD demonstrated that it is in this context of plurality and flexibility that a new national identity is being negotiated in the guise of *Ndi Umunyarwanda*. However, despite the promises of *Vision 2020*, transitional Rwanda is still dominated by the genocide of 1994 as that past which, as Fellows would have it, 'is always contemporaneous, as part of … perception of the present. In this scenario, temporality is the abrasive instrument continuously changing and layering the original narrative' (Fellows 2009: 1049).

References

Améry, Jean. 1980. *At the Mind's Limits: Contemplations by a Survivor on Auschwitz and its Realities*. Bloomington and Indianapolis: Indiana University Press.

Ansoms, An. 2009. Re-engineering rural society: The visions and ambitions of the Rwandan elite. *Journal of Modern African Studies* 46(1): 1–32.

Art for Peace. 2013. *Youth Connekt Dialogue: The Promise of a Generation. Report*. Kigali.

Baddorf, Zack. 2010. Reconciliation Village Hosts Victims, Perpetrators of Rwandan Genocide. Retrieved 3 May 2017 www.voanews.com/a/reconciliation-village-hosts-victims-prepetrators-of-rwandan-genocide-103207594/155845.html.

Bamporiki, Edouard. 2010. *Icyaha kuri bo, ikimwaro kuri jye. Igiti mwateye dushririwe n'imbuto zacyo*. Kigali.

Bar-on, Dan. 1991. *Legacy of Silence: Encounters with Children of the Third Reich*. Harvard, MA: Harvard University Press.

BBC. 2014. *Rwanda: The Untold Story*. London: BBC.

BBC Africa Gahuzamiryango. 2013. *Imvo n'imvano 1 and 2 on 22 and 29 July*. London: BBC.

Benda, Richard. 2014. BBC and Genocide in Rwanda. Conflict of Competence over Post-genocide Narrative. London: E-IR. Retrieved 3 May 2017 www.e-ir.info/2014/11/17/bbc-and-genocide-in-rwanda-conflict-of-competence-over-post-genocide-narrative/.

Berger, Alan L., and Naomi Berger (eds). 2001. *Second Generation Voices: Reflections by Children of Holocaust Survivors and Perpetrators.* Syracuse, NY: Syracuse University Press.

Burman, Erica. 2008. *Deconstructing Developmental Psychology* (revised). London: Brunner-Routledge.

Chemouni, Benjamin. 2014. Explaining the design of the Rwandan decentralisation: Elite vulnerability and the territorial repartition of power. *Journal of Eastern African Studies*, 8(2): 246–262.

Cheney, Kristen E. 2007. *Pillars of the Nation: Child Citizens and Ugandan National Development.* Chicago, IL: The University of Chicago Press.

Clark, Phil. 2010. *The Gacaca Courts, Post-Genocide Justice and Reconciliation in Rwanda: Justice without Lawyers.* Cambridge: Cambridge University Press.

Clark, Phil. 2012. *How Rwanda Judged its Genocide.* London: Africa Research Institute.

Clark, Phil. 2014. Bringing the peasants back in, again: State power and local agency in Rwanda's Gacaca courts. *Journal of Eastern African Studies* 8(2): 193–213.

Clark, Phil and Zachary D. Kaufman (eds). 2009. *After Genocide: Transitional Justice, Post-Conflict Reconstruction and Reconciliation in Rwanda and Beyond.* New York: Columbia University Press.

Collins, Randall. 2004. *Interaction Ritual Chains.* Princeton, NJ: Princeton University Press.

Crawford, Adam. 2015. Temporality in restorative justice: On time, timing and time-consciousness. *Theoretical Criminology: An International Journal* 19(4): 470–490.

Dallaire, Roméo. 2003. *Shake Hands with the Devil: The Failure of Humanity in Rwanda.* London: Random House.

Doná, Giorgia. 2012. Being young and of mixed ethnicity in Rwanda. *Forced Migration Review* 40: 16–17.

Drumbl, Mark. 2000. Punishment postgenocide: From guilt to shame to 'civis' in Rwanda. *New York University Law Review* 75(5): 1237–1253.

Dyregrov, Atle *et al.* 2000. Trauma exposure and psychological reactions to genocide among Rwandan children. *Journal of Traumatic Stress* 13(1): 3–21.

Eltringham, Nigel. 2011. The past is elsewhere: Paradoxes of proscribing ethnicity in post-genocide Rwanda. In Scott Straus and Lars Waldorf (eds), *Remaking Rwanda: State Building and Human Rights after Mass Violence*, 269–282. Madison: The University of Wisconsin Press.

Fellows, Diane. (2009). Erasure: Temporality and the second generation. *International Journal of Urban and Regional Research* 33(4): 1045–1056.

Hatzfeld, Jean. (2006). *Machete Season: The Killers in Rwanda Speak.* New York: Picador USA.

Ingelaere, Bert. 2010. Peasants, power and ethnicity: A bottom-up perspective on Rwanda's political transition. *African Affairs* 109(435): 273–292.

Kagame, Paul. (2013). *NDI UMUNYARWANDA: Isobanurampamvu kuri Gahunda ya 'Ndi umunyarwanda'.* Kigali: NURC.

Longman, Timothy. 2011. Limitations to political reform: The undemocratic nature of transition in Rwanda. In Scott Straus and Lars Waldorf (eds), *Remaking Rwanda:*

State Building and Human Rights after Mass Violence, 25–47. Madison: The University of Wisconsin Press.

Mclean Hilker, Lyndsay. 2011. Young Rwandans' narratives of the past (and present). In Scott Straus and Lars Waldorf, eds, *Remaking Rwanda: State Building and Human Rights after Mass Violence*, 316–330. Madison: The University of Wisconsin Press.

Melvern, Linda. (2000). *A People Betrayed: The Role of the West in Rwanda's Genocide.* London and New York: Zed Books.

Mugabo, Jean. 2015. How Reconciliation Villages have Fostered Reconciliation. Retrieved 3 May 2017 www.newtimes.co.rw/section/read/187661/.

Munyandamutsa, Naasson *et al.* 2012. Mental and physical health in Rwanda 14 after the genocide. *Social Psychiatry and Psychiatric Epidemiology* 47(11): 1753–1761.

Neugebauer, Richard *et al.* 2009. Post-traumatic stress reactions among Rwandan children and adolescents in the early aftermath of genocide. *International Journal of Epidemiology* 38: 1033–1045.

Newbury, Catharine. 2011. High modernism at the ground level: The Imidugudu policy in Rwanda. In Scott Straus and Lars Waldorf (eds), *Remaking Rwanda: State Building and Human Rights after Mass Violence*, 223–240. Madison: The University of Wisconsin Press.

Nsengimana, Philbert. 2014. Keynote speech on Ndi Umunyarwanda, London (notes in author's collection).

Ong, Amandas. 2017. 'My neighbour murdered nearly all my family, but now we are friends.' Retrieved 3 May 2017 www.theguardian.com/world/2017/jan/12/my-neighbour-murdered-my-family now-we-are-friends-rwanda-genocide.

Palmer, Nicola. 2015. *Courts in Conflict: Interpreting Layers of Justice in Post-Genocide Rwanda*. Oxford: Oxford University Press.

Pells, Kirrily. 2011. Building a Rwanda 'Fit for Children'. In Scott Straus and Lars Waldorf (eds), *Remaking Rwanda: State Building and Human Rights after Mass Violence*, 79–86. Madison: The University of Wisconsin Press.

Pells, Kirrily, Kristen Pontalti and Timothy P. Williams. 2014. Promising developments? Children, youth and post-genocide reconstruction under the Rwandan Patriotic Front (RPF). *Journal of Easter African Studies* 8(2): 294–310.

Peskin, Victor. (2008). *International Justice in Rwanda and the Balkans: Virtual Trials and the Struggle for State Cooperation*. Cambridge: Cambridge University Press.

Pottier, Johan. 2002. *Re-Imagining Rwanda: Conflict, Survival and Disinformation in the Late Twentieth Century*. Cambridge: Cambridge University Press.

Purdeková, Andrea. 2011. 'Even if I am not here, there are so many eyes': Surveillance and state reach in Rwanda. *The Journal of Modern African Studies* 49(3): 473–497.

Purdeková, Andrea. 2015. *Making Ubumwe: Power, State and Camps in Rwanda's Unity Building Project*. Oxford: Berghahn Books.

Rakita, Sara. 2003. *Lasting Wounds: Consequences of Genocide and War on Rwanda's Children*. London: Human Rights Watch.

Republic of Rwanda. 2012. *Rwanda Vision 2020*. Minecofin, Kigali. Retrieved 15 April 2017 www.minecofin.gov.rw/fileadmin/templates/documents/NDPR/Vision_2020_.pdf.

Reyntjens, Philip. 2004. Rwanda, ten years on: From genocide to dictatorship. *African Affairs 103*: 177–210.

Rieder, Heidi and Thomas Elbert. 2013. Lasting imprints of a genocide: Trauma, mental health and psychosocial conditions in survivors, former prisoners and their children. *Conflict and Health* 7(6): 1–13.

Schaal Susanne, Jean Pierre Dusingizemungu, Nadja Jacob, and Thomas Elbert. Rates of trauma spectrum disorders and risks of posttraumatic stress disorder in a sample of orphaned and widowed genocide survivors. *European Journal of Psychotraumatoloy* 2: 43–63.

Shearing, Clifford. 2001. Punishment and the changing face of governance. *Punishment & Society* 3(2): 203–220.

Straus, Scott, and Lars Waldorf (eds). 2011. *Remaking Rwanda. State Building and Human Rights after Mass Violence.* Madison: The University of Wisconsin Press.

Sydor, Guy, and Pierre Philippot. 1996. Conséquences psychologiques des massacres de 1994 au Rwanda. *Santé mentale au Québec* 21(1): 229–245.

Taylor, Christopher C. 2015. *Ihahamuka: An Indigenous Medical Condition among Rwandan Genocide Survivors.* Oxford: Oxford University Press.

Thompson, E. P. 1967. Time, work-discipline, and industrial capitalism. *Past & Present* 38: 56–97.

Thomson, Susan. 2011. Reeducation for reconciliation: Participant observation on Ingando. In Scott Straus and Lars Waldorf (eds), *Remaking Rwanda: State Building and Human Rights after Mass Violence*, 331–342. Madison: The University of Wisconsin Press.

Thomson, Susan. 2013. *Whispering Truth to Power: Everyday Resistance to Reconciliation in Postgenocide Rwanda.* Madison: The University of Wisconsin Press.

Umuseke online. 2013. 'Gahunda ya Ndi umunyarwanda igiye kugezwa mu nzego zose'. Retrieved 30 April 2017 http://bisituganehe.blogspot.co.uk/2013/11/gahunda-ya-ndi-umunyarwanda-igiye.html.

United Nations Secretariat. 2016. Population and Vital Statistics. Statistical Papers. *Series A*. Vol. LXVIII. New York.

Valverde, Mariana. 2014. Studying the governance of crime and Security: Space, time and jurisdiction. *Criminology & Criminal Justice* 14(4): 379–391.

Webster, Don. 2011. The uneasy relationship between the ICTR and Gacaca. In Scott Straus and Lars Waldorf (eds), *Remaking Rwanda: State Building and Human Rights after Mass Violence*, 184–193. Madison: The University of Wisconsin Press.

9 Undoing Brazil's dictatorial past

Gisele Iecker de Almeida

Introduction

Post-conflict peacebuilding takes place in messy historical circumstances, in which the past is recalled, reassessed, and, more often than not, repudiated. Attempts are made to redress or at least mitigate its effects, if not reverse the unwanted pasts altogether. As history is reworked and rewritten according to the inclinations of each historical moment, pasts have proved to be malleable and given to deconstruction and reconstruction, a re-doing and undoing promoted by 'entrepreneurs of memory' (Torpey 2009: 31): historians and educators, yes, but also human rights activists, therapists, legal professionals, documentary-makers and artists conveying the past through their art.

National identities are intrinsically linked to how the past is perceived, and thus are also a work-in-progress. This is an effect of both the reflexive character and fluidity of historical identity, and the fact that cultural memory is an active element of society, constantly circulating and in the process of being reshaped. Opening the black box of the memories of conflict in order to make modifications in the narratives sustaining national identity has the potential to help societies achieve a turning point. However, the catharsis may also unleash emotions that may have long settled, generating confusion, frustration, resentment, vengeance, regret and perhaps contrition.

Unlike critics who dismiss reparations politics by claiming that it symbolises a decline of future-oriented politics (Torpey 2006), proponents of its most influential segment, the contemporary transitional justice model (Teitel 2003), see it as a (perhaps not always very clean)[1] break 'between an abusive past and a future rights-respecting regime' (de Greiff 2010: 29); 'a process of dealing with the past in ways which assist the society to move beyond it' (Bell 2009: 19). Both de Greiff and Bell indicate that societies engaging in the activities of the politics of reconciliation and reparations seek to also address certain development and security issues, progressing as they go through the process of coming to terms with the past. Hence, past and future emerge as hermeneutic horizons taking turns in motivating societies to revisit the past. If, on the one hand, the effort of reckoning with the past entails a desire for the past to have been different and assuming a contrite attitude, on the other it also means engaging in the design of institutions capable of avoiding the future repetition of what is currently perceived as 'errors' of the past.

Evidence from the Brazilian case points at a third and yet under-researched[2] motivation to revisit a difficult past: the construction and propagation by truth and reconciliation mechanisms[3] of a historical narrative that is consistent with a desired future. This element in the politicisation of history is here described by Richard Wilson: '[p]ost-conflict governments selectively filter the past to invent a new official history and to construct a new vision of the nation' (Wilson 2005: 919). In what follows, I will analyse how the symbolic acts carried out with the intent of expressing contrition attempt to (metaphorically) undo the past in the constantly changing present. Additionally, I will present a critical perspective of the intentionality observed in this process of relieving a society of the burden of guilt by way of redrafting the national narrative. Brazil's undoing offers a practical example of how new narrative possibilities are opened. Through a revision of society's newfound perceptions of the past in light of the particular historical ambitions currently being pursued (Bal et al. 1999: 7; Gallagher 2010: 19), the bond between past and present faces deep modifications – bringing more nuance to the claim that reparations politics operates in a retrospective and prospective Janus-faced orientation (Gray 2009: 1044). Acting in the boundaries of an envisaged future, past and present, deconstruction and reconstruction (Bakiner 2015: 226), the undoing of Brazil's dictatorial past will serve as a case in point to explore reparation politics' categories of counterfactuals and historical retraction.[4]

This chapter is divided as follows: in the first section, post-authoritarian Brazil's complex historical circumstances are examined; the second part describes episodes of historical retraction in which past wrongs have been dishonoured and attempts made to reverse or mitigate their effects by undoing the past; and the third part considers the centrality of counterfactual reasoning underpinning reparation politics' dealings with the past.

Brazil's troubled past

Latin American countries that have gone through a dictatorial regime between the 1960s and 1980s can essentially be understood as sectors in the United States ideological war against the spread of communism. Deeply influenced by US intelligence agencies, generations of Latin American military officials were taught that violence was a valid and perhaps necessary instrument against revolutionary groups. Amid growing Cold War rhetoric and paranoia, states across the region terrorised their own citizens using coercive institutions and violence against the population allegedly as a tool for 'the maintenance of public order' (Verdeja 2009: 169). One way or another, many of these countries have revisited that past in an attempt to come to grips with the 'old wrongs'[5] and find a way to include the unwanted pasts in the historical narratives that forge their national identities, be it for the sake of a younger generation who have no personal recollection or first-hand experience of the period, or as a means of achieving reconciliation. After the return to democracy in Argentina, Uruguay and Chile the question of how to address past wrongs was met in each country with a disparate range of societal interest and a varying degree of power retained by the military. Argentina's recently elected president Raúl Alfonsín was the most vocal agent calling for the prosecution of perpetrators, but more characteristic of this phase of transition to democratic rule were the negotiated pacts and 'conditional amnesties (…) limiting punishment on "reconciliation grounds"' (Teitel 1995: 151). Truth commissions were held quite early on (in 1983, 1985 and 1990) – an application of the post-Nuremberg historical culture which became later known as transitional justice (Teitel 2003). Based on legal discourse and litigation procedures, transitional justice scholars and practitioners have designed instruments for use when past conflicts are the source of political struggles in the present.

It would be several years before the Brazilian state would disown the 1964–1985 dictatorial regime and repudiate the crimes against humanity perpetrated on its behalf. Brazil's authoritarian military regime was marked by the curtailment of citizens' political and human rights and the restriction of civil liberties. Political opposition or even criticism of the regime was met with brutality from the security forces, with disappearances, illegal killings and torture used against those suspected of being dissidents. What started with a bloodless coup and eventually became a brutal regime, ended with a negotiated handover of power. The country's complicated process of political transition was not initiated until the 1979 self-amnesty (a central element in the military option for a 'slow, gradual and sure relaxation'[6]) was put in place, ensuring to this day the impunity of crimes committed by state agents. Although millions of Brazilians protested in favour of direct elections in 1984, the military staved off *Diretas Já* and an indirect election[7] was held to form the New Republic of 1985, de facto an interim government during which a new constitution was drafted and crimes committed by the military became taboo. According to historian Carlos Fico, the nation grew deeply frustrated as the prolonged and complex process of military step down and handover of power unfolded (Fico 2013). The silence about the crimes committed by representatives of the dictatorial regime meant they were swept under the rug and overlooked, in a process remarkably similar to Robert Gordon's description of collective forgetting:

> willed oblivion is sometimes the most conservative story of all: it brackets the period of injustice as a historical accident, an outlying event in the general stream, so atypical as not to be worth mentioning, with no origins traceable to the prior period or implications for the present one. The injustice epoch drops out of memory altogether, and history resumes at the point where it left the track.
>
> (Gordon 1996: 40)

The choice to leave the past behind meant the entrenched sectarianism of the radicalised extremities of the political spectrum was never resolved, and with it the divergence of opinion over whether those tortured and killed by the military should be deemed national heroes or *terroristas* was never addressed. For the majority of the Brazilian population who neither opposed nor endorsed the regime, there is often an enormous confusion as to what had unfolded and how the past should be perceived. Effective military censorship at the time and the fact that the period was not often mentioned in classrooms (Pereira and Pereira 2011) meant that many Brazilians could – and many still do – live blissfully unaware of the fact that the state was responsible for heinous crimes against its citizens just a few decades ago. During Brazil's silent period, conflicting perspectives about the past solidified, myths and stereotypes were created and a simplistic, dualistic view of heroes vs. villains was reinforced (Fico 2004). An example is memoirs written by former militants, some of which put forward a conciliatory leftist memory and a somewhat romanticised view of the dictatorial period. Fernando Gabeira's bestselling first-person account *What's Up, Comrade?* published in 1979 is said to have had a particularly long-lasting effect in Brazilian historical consciousness (Rollemberg 2006), enforcing the view that the radicalisation of the left was purely naïve (Fico 2012, 25), and therefore not dangerous or criminal.

In time, a heroicisation of the left has taken hold (Fico 2012, 29); ignoring the fact that most armed leftist guerrilla groups sought to implement a (not particularly democratic) socialist regime, the counterfactual narrative which portrays the radical left as some sort of heroic democratic resistance, a last democratic bastion and their arms the last resort of the righteous in the fight against the military gained prominence (Reis 2014). This perspective made a 'tweak' in the narrative possible; it enabled a modification in Brazil's representation of its past. The portrayal of the military as thugs and the leftist militants as armed defenders of Brazilian democracy meant the violence during the regime could be memorialised as a war, and society could align itself with the 'good side', that is, of course, in favour of democracy (Reis 2010; Huyssen 2005). Because the blame was put at the door of the military barracks, the slate was wiped clean on big business support and financial gain from the regime, as well as civil society's conformity and in many cases cooperation with the military. Just like that, the brunt of society's share of responsibility for the events of the past was omitted. This is how, much to the resentment of members of the armed forces, Brazil's swing to the right, quite popular at the time, became largely perceived as a low point in the country's past (D'Araujo 2014). The tide of history had reached a turning point.

Underpinning Brazil's present, the narrative fits nicely alongside the country national identity – a racially mixed country where its non-violent people coexist peacefully, a human rights-defending society that, in line with its long-held democratic values, endured and overcame the dictatorial regime (Chauí 2013; Reis 2010). And, more importantly for us, it speaks of the future Brazil wants to convey, namely its ambition to be acknowledged in the international arena as a country that promotes political correctness and upholds international human rights norms and discourse (Macaulay 2010). Unchallenged within South America, Brazil now seeks to be perceived as an equal among civilised nations – an ideal reflected in its increasing dedication to solving international security crises and made palpable in efforts to secure a permanent seat at the United Nations Security Council. To carry out 'the duty of lifting the lid off painful periods of history', which Brigitte Weiffen argues 'has almost become an international norm' (Weiffen 2012: 89), is an effort that can only enhance Brazil's chances of reaching that goal. All past misconduct must be dealt with: no more skeletons in the closet.

Important as Brazil's interest in becoming a full member the global community might have been, a domestic shift in the political sphere played a pivotal role in tipping the country towards its overdue reconciliation process and belated engagement with the memory of the dictatorship. After 2005 in particular, many individuals who stood against the regime in their youth and later pursued political careers, reached the upper echelons of power (Schneider 2011). A former political exile, a former prisoner and a former member of the armed opposition during the regime, presidents Cardoso, da Silva and Rousseff, have all risen to become Brazil's commanders-in-chief. It really should come a no surprise if signs of an alignment with the leftist perspective can be identified in Brazil's 'politics of reconciliation by memory' (Atencio 2014: 123). Either way, it became increasingly clear that the country had moved with the times, and, albeit much later than its neighbours, Brazil's executive and legislative powers launched projects to repair the wrongs of the past. In this process, the state took responsibility for the crimes committed by the armed forces, offered moral and financial compensation to victims and sought to establish a new understanding of the years of violence. The projects making up Brazil's large but uncoordinated reparations programme, have so far excluded the criminal prosecution of perpetrators and distributive justice, focusing instead on financial and moral reparations alongside truth initiatives. Beyond compensation to individual victims, their activities include documentation-gathering projects, the publication of history books, the construction of new memorial sites and a truth commission.[8]

In Brazil's case the establishment of a historical record, or elaboration on the past, was obtained first through investigations conducted by victims and their families. Only years later would a state-commissioned research be launched to build on the early efforts of victim groups and investigate the gross violations of human rights and crimes against humanity perpetrated by state officials. The 2007 report *Right to Memory* presents a national narrative that can be perceived as a public *mea culpa* by the state, an 'I did it' which, at the same time, also accuses the armed forces of wrongdoing (a 'they did it'). It would seem a 'we did it' account, emphasising society's support for the regime is not forthcoming.[9]

Altered pasts

This section focuses on some of the ways in which the past can be undone, that is, dealt with in such a way that past wrongs are rolled back, its effects reversed and a new narrative that stitches together the society's repudiated pasts and ambitioned future becomes more established. Undoing is the attempt to make it seem *as if* the past had never happened by creating different timelines which allow the adoption of an alternative future; 'to change the present by subtracting a crucial past event and thereby sending history off in a new direction' (Gallagher 2002: 11). Perhaps undoing can be understood as a process of mediation of the many pasts in existence, not only all those options avoided or those 'unrealised pasts', which might still be 'alive within us' (Gallagher 2002: 24), but also 'memories that people – and even historians – have chosen not to live with' (Southgate 2007: 187).

Brazil's reformed views of the dictatorial regime generated attempts from politicians at all levels of government to demonstrate they were on the right side of history, so to speak. The reputation of individuals honoured for their service to the regime might now be deemed unworthy of tribute and these honours withdrawn, as occurs in the tearing down of statues of high-ranking military authorities and the renaming of public spaces once named after them. This type of attempts to repair the present's perception of the past carries a distinct will to move towards a new representation of the past: out with the old, in with the new. One criticism proponents of change will undoubtedly encounter is that such acts may constitute an exercise in forgetting (Ricoeur 2006), an institutional cleansing of a nation's ugly bits. Strategies to counteract the claim and potentially empower the public to enquire about the historical period and the circumstances that demanded the changes include selecting a new name that is thematically connected to the old one and memorialising other aspects of the same era (Yale University 2016). Brazilian public spaces, for example, are often renamed after individuals who opposed the regime, and most newly built monuments commemorate its victims.[10] I agree with David Lowenthal, who claims that '... we should be wary not only of condemning but of eternalizing ancestral evils' (Lowenthal 2009: 965). When performing an intervention in the memory of its seemingly unwanted past, societies engage as much in the construction of the national identity and the future it wishes to forge for itself as in the deconstruction of a legacy and memory thought to be fundamentally at odds with that vision.

One such person on Brazil's list of *personae non gratae* is its most ruthless dictator, Arthur da Costa e Silva. His fall from grace has followed a wide debate over whom to commemorate and honour – and how to put right the wrongs of the past. Recently, Sao Paulo and Brasilia have decided to eradicate his name from city maps, and Rio de Janeiro has a draft bill and should follow suit. In all three cities, the change of designation has resulted in the memorialisation of individuals who were persecuted by the regime.[11] Also, only a few days after the publication of the Brazilian National Truth

Commission's final report which attributed responsibility for crimes against humanity to the dictator, a bust erected in his honour in 1974 in his hometown of Taquari was torn down. Amid some polemic and debate in the social media, the mayor declared the city would 'no longer pay homage to a person who (...) does not deserve to be honoured' (Panke 2014).

The Brazilian National Congress is another institution that has shown to be very keen to shake off its cooperation with the military – especially due to a session held in 1964 in which the leader of the Senate declared the presidency vacant, paving the way for the *coup d'état*. One attempt to correct memory and redirect society back to normality following the state of exception, putting history back on track to return to Gordon's metaphor, has been the devolution of the mandates of members of congress expelled from office by the regime. With only one vote against the motion, the symbolic gesture sent the signal that in 2012 the Congress thought its role in the coup had been a regrettable mistake, something that should never have happened. An even more awkward retraction took place just a year later; in what the leader of the Senate judged to be 'an opportunity to repair History and (...) repair the constitutional role of Parliament', the 1964 declaration that had led to the end of President João Goulart's presidency was annulled. Effectively, it was *as if* the present Congress had turned back time and corrected its entry in the annals of history; now once again deemed worthy of a full state apology, it was *as if* Goulart, who had been ostracised and died in exile, had never been ousted. As if the coup had been a wrong turn that had to be dealt with, so that the country could (finally) get back on its right course. Emblematically, the day Goulart's exhumed remains were paraded with full state honours in Brasilia was declared by president Rousseff 'the day Brazil encountered its history'. Albeit symbolically, history was recalibrated.

The posthumous retroactive devolution of President Goulart's mandate is not the only instance in which history has been rewritten so as to flatter the present; neither is Brazil the only country in which these Janus-faced forward-looking symbolic retrospective gestures are taking place. Take for instance France's Gayssot Act (which punishes Holocaust denial), the wide-ranging Historical Memory Law challenging the *Pacto del Olvido* in Spain, or even the students-led calls for the statue of the British colonialist Cecil Rhodes to be removed from the campus of the University of Cape Town as in their view 'it promoted institutionalized racism and promoted a culture of exclusion particularly for black students' (Bosch 2017). The campaign crossed national borders and at Oxford the heated debate lead to a decision to keep the statue in its place to 'help draw attention to this history' (Oriel College, Oxford University 2016). Again, in Britain, a posthumous royal pardon was issued to early computing developer Alan Turing. Convicted of gross indecency in 1952, the war hero ended his own life just a few years later. Under 'Turing's Law,' now that homosexual relations have ceased to be a criminal matter, thousands of men convicted for sexual offences in the past can request that their convictions be erased from their records in what is known as a disregard process. The episode illustrates that most symbolic retrospective gestures 'cost legislators nothing, mean nothing, commit them to nothing and are designed to improve only one thing – their reputations' (Bennett 2016). Indeed, memory is mobilised and the past re-elaborated not for the sake of the past itself, but to allow the present to claim the moral high ground, as if to say, 'mistakes were made in the past, but now we see that was wrong, that is no longer part of who we are'. This realisation, or 'change of heart' and obliteration of a particularly painful, embarrassing or humiliating past folly is the lynchpin upon which a new understanding of the past with the potential of 'sending history off in a new direction' can be forged (Gallagher 2002: 11).

The process of redrafting Brazil's history by renaming and shaming remains unfinished, with a number of schools, hospitals, streets and squares honouring fallen heroes still waiting for their turn to be renamed. Another reversal yet to occur is the withdrawal of military honours due to 'relevant service for the nation and the armed forces' granted when those 'services' (which included crimes against humanity) were part of the status quo. Scholar Fábio Comparato is unwilling to accept that the honours bestowed upon Dalmo Lúcio Muniz Cyrillo, for example, have so far not been revoked (Comparato 2001: 36). In a twisted turn of events, the infamous torturer received a peace medal in 1973 ('*Medalha do Pacificador*') and a Brazilian Knighthood Honour during Itamar Franco's presidency, twenty years later. Cyrillo's honours are an unfortunate way of highlighting that 'the file of memory is never closed; it can always be reopened and reconstructed with new acts of remembering' (Assmann and Shortt 2012: 3), and that society must remain vigilant to historical injustice.

Outrageous as a democratic government honouring perpetrators might sound, it is important to differentiate changes in public opinion and tweaks or modifications in a society's perceptions of the past from blatant forgery and deception, as for example occurs with the 'Riocentro' case, as is known the episode in which military officials conspired to plant an exploding device under the stage of a crowded concert hall in Rio de Janeiro, with the intention to blame 'leftist terrorists' (as the military would call members of the leftist guerrilla) for the blast. The carnage never happened, because the bomb exploded outside the Riocentro compound, killing and severely injuring the officials responsible for its transportation (Brasil 2014b). A military inquiry held in 1988 was adjourned with the allegation that the crime, committed in 1981, was inscribed in the 1979 amnesty law. This means that an attempt against the lives of thousands of civilians was retroactively declared pre-pardoned. Rather than an expression of contrition and symbolic undoing of history, the military's fiddling with the past was a blatant act of retrospective opportunism: a manipulation of memory and rewriting of history comparable to the historical revisionism practised by the dystopian Ministry of Truth in Orwell's *1984*: Brazil's very real reminder that 'he who controls the present controls the past'.

Alternate histories

To this day Cold War rhetoric and anti-communist sentiment has continued to influence Brazilians' popular psyche, with the Cuban revolution occupying a very prominent role in the retained symbolism of that era. Continued fears that 'Brazil is becoming a large Cuba' or that it might be 'following Cuba's footsteps', and expressions such as 'Go back to Cuba!' – chanted as recently as in last year's public manifestations in favour of Rousseff's impeachment procedure[12] – demonstrate the significant impact of Fidel's revolution. Similar reasoning can be perceived in a widely held justification for the dictatorial regime that claims it was the lesser of two evils: *if* the military had not taken power, Brazil *would have* adopted a Cuban-style communist regime. These ideas are an example of retroactive projections and counter-historical approaches also known as counterfactuals, that is, an exploration of scenarios based on what did not happen in the past but could potentially have done.[13]

Measures such as those explored in the previous section, taken with the intention of undoing, rolling back an unwanted or regrettable past in an attempt to bring about an ambitioned future, also rely on counterfactual reasoning, though with a slight twist. Catherine Gallagher has explored the workings of alternative history narratives and historical justice, and identifies two ways by which reparations can be counterfactual, one is along the lines of old-school critical military history, which seeks to answer forward-looking questions such as 'How can we fight better wars?' and the second, a backward-looking type of reparations counterfactuals, which 'pursued justice by the specific means of *undoing* some state of affairs brought about by the previous governmental actions' (Gallagher 2010: 19). Gallagher also points out that 'what if … ?' scenarios have the potential to establish some of the alternative actions and policies that were available at the time, making them a useful resource in ascertaining the culpability by inaction of certain agents and groups in society.

The examples presented above taken from the Brazilian case and elsewhere, suggest that in these alternate histories arising from atonement the 'what if' question is posed in the negative: '*what if* X had never happened?' (in which X is the error present measures try to obliterate). Take Goulart's case, for instance. The final result of the posthumous devolution of his presidential mandate was that he could be reinterred *as if* he had *never* been ousted by Congress. Residents of Taquari can now visit their square and feel *as if* a dictator had *never* been honoured by their fellow *taquarenses.* Once Cyrillo's medals are withdrawn, Brazilians will be able to feel relieved that a historical wrong has been made right, because from then onwards it will be *as if* he had *never* been honoured.

Attempts to carry on *as if* certain regrettable things could be undone are indicative of reparations' future and past orientation: to backtrack on the past and withdraw from a previously chosen historical path whilst shaping a different future through symbolic retrospective gestures. Luis Gutiérrez hit the nail right on the head in explaining that he would not attend the forty-fifth US presidential inauguration because to attend would be *as if* 'everything that candidate Donald Trump had said (...) is OK or erased from (...) memory' – things that the Democrat claims he 'can never un-hear' (Gutiérrez 2017). The Congressman's speech implicitly condemns attendees who were as appalled as he was by Trump's declarations, but rather than remain in their previously chosen historical path and continue to resist the Republican, had supposedly chosen to 'un-hear' Trump's inflammatory remarks, come to terms with and normalise the fact that he was to become US president.

Thus, the present's recovery and 'use' of altered pasts exposes the 'interpretive fluidity of history' (Weyeneth 2001: 38), the fact that at each historical moment, the past is perceived anew. Hungarian philosopher Agnes Heller draws our attention to the connection between the revisability of the past and national identities, pointing out that aspects picked up today are a selection of those things from the past (symbols and values) that have a meaning for the moment we currently live – things that we perceive as the origin of, or related to, who we are today (Heller 1982). Tomorrow, next week, 10 or 100 years from now, something else will be the object of a revision, other pasts will be made to re-emerge and other futures will be unlocked. The Victorian convention of treating children as adults is a useful example; as much as we would like to think that the present is some sort of apex of civilisation (as many Victorians did), decisions made then, much like the decisions made today, are at the mercy of a continuous revision process: what we take as normal might well be reconsidered in the future, both in light of forthcoming developments (unknown to us at the moment), or due to a reading of the present status quo from a perspective or norms that are different from ours (say for example in accordance to new established moral views or political aspirations). Or have we not reached an unspoken consensus that it is not acceptable to send children to hard labour prisons when they commit a crime?

Certain symbolic retrospective gestures and other measures to right past wrongs that can be said to attempt to undo the past and project a 'better' future, appear to be enmeshed in a concept of time in which temporal categories are perceived as interlinked and engaged in a formative relation that is contingent and continuously in flux.[14] Having said this, the process of modifying the past under consideration is not an outlandish attempt to travel in time or some quixotic backward causation project. It is rather simpler than that: we are looking at one of the ways in which, at any given present time, a perception of the past can be changed in order to make it seem as if it had never happened.

Somewhat more elaborately put: a diverse temporal dynamic emerges in periods of national redefinition in which societies work through a regrettable past, a past that is no longer agreeable with a lived present and ambitioned future. Societies in their historical present may tap into the gigantic depository of possibilities of their unrealised historical pasts, and, by provoking a deviation in the sequential ordering of historical developments, feed a preferred unrealised past into the present, forging a new nexus that opens a (future) possibility that had not existed until that moment and bringing to light a complex pluridimensional, non-linear temporal relation between past, present and future – here scrutinised through an assessment of post-authoritarian Brazil's attempts to undo past decisions as part of a larger effort to deal with its difficult past.

Although it is not possible to 'un-chop a tree' (Verdeja 2009), contemporary Brazil's attempts to reach some form of reconciliation and striving to deal with its objectionable past, present us with some interesting examples of undoing of the past through symbolic retrospective gestures and mobilising the past to remedy historical injustice. The term 'tweak' is most appropriate for the process described here, as it highlights a certain awareness of the wrongdoing and expresses the intentionality of the process of re-imagining and bringing to life historical alternatives that did not unfold, attaching the present to a different timeline. This is a deliberate act of self-improvement, more often than not an interested reassuring reminder that creates a distance between present and past and stabilises the current status quo. However, this use of the past is a dangerous resource when employed in polarised societies that remain divided by the past conflict and still find themselves in the process of negotiating its historical narrative and planning their future and how they intend to live together. The challenges of reaching a consensus over a shared representation of the past are quite clear, and undoing ultimately has the potential to hamper society's efforts to overcome those divisions, as it tends to simply gloss over the problems and promote forgetting rather than promote reflection and understanding of the events of the past.

Conclusion

The debate about the past and future horizons motivating societies to revisit a difficult past has so far been primarily framed within the aspects of engaging in a process of reckoning with the past, and strengthening institutions to avoid repetition of errors in the future. However, this approach seems inadequate to fully understand the thus far under-theorised balancing act between the past and future horizons of reconciliation projects. An example is the complex pluri-dimensional, non-linear temporal relation between past, present and future at play in attempts to undo the past, a perilous process, perhaps, though it would seem one that is widely used to unburden a society from a remembered past that has ceased to be regarded as fitting alongside a historical narrative that is consistent with a society's desired future.

This chapter offered a critical perspective of undoing, portrayed as a deliberate act of self-improvement that lays bare the intentionality of the process of inventing a new official history by revisiting an unwanted past in order to reshape, review and redraft a society's perspective of regrettable occurrences thus constructing a new vision of the future. The fact that Brazil launched a wide-ranging (if uncoordinated) reparations programme just as the country was forging a new identity as a significant player in the international system and as politicians with a particular view on the events of the past gained prominent seats in Brasilia has not gone unnoticed. My approach, drawing on an analysis of some of the undoings identified within the Brazilian process of reckoning with the human rights violations perpetrated during its authoritarian regime, has been to expose the visible hand operating the controls, disclosing that, as one commentator put it, 'we have become conscious (…) that someone must choose – omitting, combining, rearranging the details of the past in an active way – the stories by which we develop our collective memory and define our national identity' (Osiel 2012: 237).

Although going back in time and creating different timelines may seem conceptually quite close to forgetting or 'willed oblivion', they are not the same phenomenon. For example, with the renaming of public spaces, as we have seen in Brazil: to rename a space that once commemorated a dictator by completely changing the name for something that has no connection to the past would constitute an exercise in forgetting, an erasure of the past. But to withdraw from that initial homage, and rename that same space in honour of a victim for example, would constitute what we have called here an attempt to undo the past – the operation of tweaking the past in order to promoting a perspective of the past that supports a preferred vision of the future. That backtracking, the rolling back, the change of mind and the choice of following another path from now on, in our example, a path in which dictators are not dignified with the honour of naming public spaces, is the distinct trait of undoing. The analysis of attempts to (metaphorically) undo the past has brought to light the fact that this particular way of dealing with the past often does not lead to meaningful social or political transformation. In Brazil's case, it is widely known that authoritarian practices continue to be normalised in society. The widespread incidents of police violence and overcrowded detention facilities are reported to be of deep concern by the UN Working Group reviewing Brazil's human rights record in 2017.[15]

The act of revisiting the past and reshaping how it is perceived in the present in line with a more desirable future then comes across as a political trickery that is more concerned with the legitimation of current political agents or ideas rather than addressing or correcting past injustice. Rather than a principled attempt to do the right thing and 'handle it justly' (Chen and Kung 2015) as politicians and transitional justice enthusiasts might claim, it is often the case that policy makers act out of their own self-interest, be it to give the impression of being on the right side of history, which allows the present to take the moral high ground, or even to appease the international community. Often, revisiting the past leads to a narrative that serves as a reassuring reminder that stabilises the present status quo. One could say in Whitean terms that conflict-torn societies are busy rejigging the employment of their identity-forming narrative from a tale of destruction into a tale of redemptive triumph or harmonious reconciliation (White 1985), but in Brazil that process is incomplete.

Brazil's transition to a positive identity-forming narrative of redemptive triumph or harmonious reconciliation at one point seemed feasible, especially around 2010 when the country was being pictured internationally as a global power on the rise. Brazil had improved the living conditions of the poorest in society, it was scheduled to hold major international sports events and it looked as if it would have the opportunity to play a leading role in an increasingly multilateral world order. At present, though, it looks as if the sweet spot the country had carved for itself has been blighted by financial instability, corruption scandals, the outbreak of a major health crisis and increased sectarianism following the conclusion of its second impeachment process in less than two decades. Crucially, the appalling state of human rights cannot be easily glossed over. Living in some of the world's most dangerous cities,[16] Brazilians have grown used to police violence and recurring authoritarian practices such as those denounced in the truth commission report. How can it be possible to draw a line that differentiates the past and the future and establish a new narrative breaking away from the violent past in the absence of significant political transformation?

Even though the historical ambitions Brazil at one point pursued now look further away, there is no end in sight to its expression of contrition and use of reparation counterfactuals in symbolic measures to right past wrongs. Caught in a process of constant reinvention, Brazil uses the things it regrets having done in the past to project the kind of society it would like to become. Or, to paraphrase an expression coined by Seymour Mandelbaum, Brazil is choosing its past while it chooses its future (Mandelbaum 1977: 193).

Notes

1 Jorge Heine claims 'nations need a moment of 'rebirth' (…) one that a clean break with that oppressive past and thus (…) the necessary impetus to forge ahead and build a better future' (Heine 2007: 72).

2 One author who has consistently analysed the narrative element of political transitions is Ruti Teitel. See her chapter on 'Historical Justice' in *Transitional Justice* (2000), and the reprinted papers in the section 'Narratives' of her volume *Globalizing Transitional Justice* (2014). For an analysis of narratives produced in truth commissions in light of Hayden White's theories see Buckley-Zistel, Susanne. 2014. 'Narrative Truth: On the Construction of Truth in Truth Commissions.' In *Transitional Justice Theories*, edited by Susanne Buckley-Zistel et al., London: Routledge.

3 My research focuses specifically on the initiatives launched by the executive branch of the Brazilian state. Post-conflict efforts have so far included truth initiatives (such as the release of official historical reports, new archives and a truth commission) as well as projects towards reparations to victims (including the public recognition of wrongdoing, memorials, financial restitution for certain losses and other forms of symbolic redress). No perpetrator has been prosecuted, in accordance to the 1979 Amnesty Law.

4 Two complementary mappings of the field inform this view: one is Will Kymlicka and Bashir Bashir's notion of politics of reconciliation as consisting of debates on the ethics of apology; forgiveness; trauma and healing; theories of collective

responsibility; memory and remembrance; and reparations (Kymlicka and Bashir 2008: 6). The other is John Torpey's diagram of reparations politics (Torpey 2003: 6) in which transitional justice's policies consist of an element within the larger field of reparations.

5 Old wrongs appear between inverted commas to indicate that very often the problems that arise when the past is reassessed stem from the fact that the events of the past are not deemed 'wrong' or 'evil' by everyone. A large part of the population and the army officials who perpetrated crimes on behalf of the state continue to claim the use of violence was necessary, and they did nothing wrong. The more recent rise of retrospective politics and transitional justice can be understood in light of changing political mores and conceptions about what is politically or socially legitimate and what is wrong (certainly with regards to the use of violence).

6 These are the words famously used by dictator General Ernesto Geisel in 1974 to describe the political transition he would start. Indeed, it was long, as Brazil's first presidential elections post-dictatorship would only be held 15 years later.

7 The Brazilian dictatorial regime maintained what some commentators consider an 'appearance of democracy' or a 'democratic façade' – that is, certain democratic elements with authoritarian traits, such as 'bionic' members of Congress who would vote according to the military regime's whip. The 1985 'indirect' presidential election was the final act in this tradition; it was the system through which members of Congress elected the first civilian president since 1964, putting an end to military rule in Brazil.

8 The Brazilian National Truth Commission gave way to over 100 local commissions established by local authorities or NGOs across the country. Most of these commissions supported the federal commission and their work informed its final report. For a more thorough analysis of Brazil's 'National Truth Commission' see Schneider and de Almeida, forthcoming.

9 As an example of this propensity towards an 'I/t National Truth Commission's final report (published in 2014), the regime is referred to as a 'military dictatorship,' even though the term 'civil-military dictatorship' is increasingly used by Brazilian intellectuals precisely to highlight the level of support the regime had at the time. The latter terminology appears in the report's second tome, which consists of outsourced chapters (Brasil 2014a; D'Araujo 2015). For a discussion on the use of pronouns, see Meister 2011.

10 Countless new monuments, memorials and busts have been constructed commemorating the militants who opposed the regime. Examples are found in Recife, Sao Paulo, Rio de Janeiro, Porto Alegre and Belo Horizonte, where the 'Memorial da Anistia,' a museum also in honour of the victims of the regime, is expected to be built.

11 In 2016 the 'Elevado Presidente Costa e Silva' (in Sao Paulo) has been renamed 'Via Elevada Presidente João Goulart.' Brasilia's renaming of the Costa e Silva bridge in honour of student leader Honestino Guimarães was the last step in the purging of public spaces in the country's capital named after military leaders, a process that had begun in 1985. In Rio de Janeiro the proposed law changes the Rio-Niteroi 'Costa e Silva' bridge's name to 'Herbert (Betinho) de Souza' bridge, after the well-known sociologist, who was a political exile during the regime and a devoted champion of human rights who fought for the eradication of hunger in the country upon his return.

12 The expression 'Volta pra Cuba!' is used in Brazil as an insult against left-leaning political adversaries. It denotes that an idea or person is considered more suitable for the Caribbean socialist state where members of the armed opposition (such as Rousseff herself) received military training in the 1960s.

13 Jon Elster dedicated an entire chapter of his canonical *Closing the Books* (Cambridge University Press, 2004) to counterfactual reasoning in justifications and excuses for wrongdoing.

14 I am thinking in particular of Edgar Morin's formulation of a complex temporality: 'there is always a retroactive game between past and present in which not only does the past contribute to the knowledge of the present, which is obvious, but also the experience of the present contributes to the knowledge of the past, thus transforming it' (Morin 1986: 304). See Zeitlyn 2015 for a survey of alternatives and a defence of 'pluralising' the past.

15 Brazil's third Universal Periodic Review is ongoing. A preliminary report can be consulted at: https://documents-dds-ny.un.org/doc/UNDOC/GEN/G17/045/78/PDF/G1704578.pdf?OpenElement.

16 Nineteen of the world's fifty most violent cities are located in Brazil according to the 2015 report of the non-governmental organisation *Consejo Ciudadano para la Seguridad Pública y la Justicia Penal.* Four cities appear in the ranking's top ten: João Pessoa, Maceió, Fortaleza and São Luís (CCSPJP 2015).

References

Assmann, Aleida and Linda Shortt (eds). 2012. *Memory and Political Change.* New York: Palgrave Macmillan.

Atencio, Rebecca J. 2014. *Memory's Turn: Reckoning with Dictatorship in Brazil.* Madison: University of Wisconsin Press.

Bakiner, Onur. 2015. *Truth Commissions: Memory, Power, and Legitimacy.* Philadelphia: University of Pennsylvania Press.

Bal, Mieke, Jonathan V. Crewe and Leo Spitzer. 1999. *Acts of Memory: Cultural Recall in the Present.* Hanover, NH: University Press of New England.

Bell, Christine. 2009. Transitional Justice, Interdisciplinarity and the State of the 'Field' or 'Non-Field'. *International Journal of Transitional Justice* 3(1): 5–27.

Bennett, Catherine. 2016. Pardoning the Dead Is Fine. Better to Say Sorry to the Living. *The Guardian*, 23 October, sec. Opinion. Retrieved 4 June 2017 www.theguardian.com/commentisfree/2016/oct/23/alan-turing-posthumous-pardons-gay-lgbt-rights.

Bevernage, Berber. 2011. *History, Memory, and State-Sponsored Violence: Time and Justice.* Routledge Approaches to History. New York, London: Routledge.

Bosch, Tanja. Twitter Activism and Youth in South Africa: The Case of #RhodesMustFall. *Information, Communication & Society* 20(2): 2017.

Brasil Comissão Nacional da Verdade. 2014a. *Relatório / Comissão Nacional Da Verdade / Tome I.* Brasilia: CNV.

Brasil Comissão Nacional da Verdade. 2014b. *Relatório Preliminar de Pesquisa Caso Riocentro: Terrorismo de Estado Contra a População Brasileira.* Brasilia: CNV.

Chauí, Marilena. 2013. *Manifestações Ideológicas do Autoritarismo Brasileiro.* Belo Horizonte: Editora Autêntica.

Chen, Jau-hwa and Kung, Wei-cheng. 2015. The Grammar of Transitional Injustice. Paper presented at the symposium Transitional Justice – The Role of Historical Narrative in Times of Transitions held at Heidelberg University.

Comparato, Fábio Konder. 2001. Ética política e honra militar. In Janaína Teles (ed.), *Mortos e desaparecidos políticos: reparação ou impunidade?* Sao Paulo: Humanitas FFLCH/USP, 55–63.

Consejo Ciudadano para la Seguridad Pública y la Justicia Penal. 2015. The 50 Most Violent Cities in the World. Retrieved 4 June 2017 www.seguridadjusticiaypaz.org.mx/biblioteca/prensa/send/6-prensa/199-the-50-most-violent-cities-in-the-world-2014.

D'Araujo, Maria Celina. 2014. Fifty Years since the Military Coup: Taking Stock of Brazilian Democracy. CMI – Chr. Michelsen Institute. Retrieved 4 June 2017 www.cmi.no/publications/5225-fifty-years-since-the-military-coup.

D'Araujo, Maria Celina. 2015. Taking Stock (with Discomfort) of the Military Dictatorship Fifty Years after the 1964 Coup: A Bibliographical Essay. *Brazilian Political Science Review* 9(3): 143–163.

Fico, Carlos. 2004. Versões e controvérsias sobre 1964 e a ditadura militar. *Revista Brasileira de História* 24(47): 29–60.

Fico, Carlos. 2012. Brasil: a transição inconclusa. In Maria PaulaN. Araujo, Carlos Fico, and Monica Grin, eds., *Violência na história: memória, trauma e reparação.* Rio de Janeiro: Ponteio, 25–37.

Fico, Carlos. 2013. Violence, Trauma, and Frustration in Brazil and Argentina: The Role of the Historian. *Topoi* 14(27): 239–261.

Gallagher, Catherine. 2002. Undoing. In Karen Newman, Jay Clayton, and Marianne Hirsch (eds), *Time and the Literary.* New York, London: Routledge, 11–29.

Gallagher, Catherine. 2010. Telling It Like It Wasn't. *Pacific Coast Philology* 45: 12–25.

Gordon, Robert W. 1996. Undoing Historical Injustice. In Austin Sarat and Thomas, R. Kearns (eds), *Justice and Injustice in Law and Legal Theory.* Ann Arbor: University of Michigan Press, 35–75.

Gray, David C. 2009. A No-Excuse Approach to Transitional Justice: Reparations as Tools of Extraordinary Justice. *Washington University Law Review* 87: 1043–1104.

de Greiff, Pablo. 2010. A Normative Conception of Transitional Justice. *Politorbis* 50: 17–29.

Gutiérrez, Luis V. 2017. *Why I Will Not Be at Inauguration and Will Be Marching with Women.* Transcript, Gutiérrez Media Center, 10 January. Retrieved 15 January 2017 https://gutierrez.house.gov/media-center/press-releases/why-i-will-not-be-inauguration-and-will-be-marching-women.

Heine, Jorge. 2007. All the Truth but Only Some Justice? Dilemmas of Dealing with the Past in New Democracies. In Edel Hughes, Ramesh Thakur and William A. Schabas (eds), *Atrocities and International Accountability.* UN Publications, 65–80.

Heller, Agnes. 1982. *A Theory of History.* London: Routledge.

Huyssen, Andreas. 2005. Resistance to Memory. In M. Pensky (ed.), *Globalizing Critical Theory.* Lanham, MD: Rowman & Littlefield, 165–184.

Kymlicka, Will, and Bashir Bashir. 2008. Struggles for Inclusion and Reconciliation in Modern Democracies In *The Politics of Reconciliation in Multicultural Societies.* Oxford and New York: Oxford University Press, 1–24.

Lowenthal, David. 2009. On Arraigning Ancestors: A Critique of Historical Contrition. *North Carolina Law Review* 87(3): 901–966.

Macaulay, Fiona. 2010. Human Rights in Context: Brazil. In Mónica Serrano and Vesselin Popovski (eds), *Human Rights Regimes in the Americas.* Tokyo: United Nations University Press, 133–156.

Mandelbaum, Seymour. 1977. The Past in Service to the Future. *Journal of Social History* 11(2): 193–205.

Meister, Robert. 2011. *After Evil: A Politics of Human Rights.* New York: Columbia University Press.

Morin, Edgar. 1986. *Para sair do século XX.* Rio de Janeiro: Nova Fronteira.

Oriel College, Oxford University. 2016. 'Statement from Oriel College on 28th January 2016 Regarding the College's Decision Concerning the Rhodes Statue'. Retrieved 4

June 2017 http://production.oriel.ox.ac.uk/sites/default/files/statement_from_oriel_college_on_28th_january_2016_regarding_the_college.pdf.

Osiel, Mark. 2012. *Mass Atrocity, Collective Memory, and the Law.* New Brunswick, NJ: Transaction Publishers.

Panke, Marcela. 2014. Remoção de busto do ex-presidente Costa e Silva causa polêmica em Taquari. Gaúcha. Retrieved 4 June 2017 http://gaucha.clicrbs.com.br/rs/noticia-aberta/remocao-de-busto-do-ex-presidente-costa-e-silva-causa-polemica-em-taquari-125376.html.

Pereira, Mateus, and Andreza Pereira. 2011. Os Sentidos Do Golpe de 1964 Nos Livros Didáticos de História (1970–2000): Entre Continuidades E Descontinuidades. *Tempo* 16: 197–220.

Reis, Daniel Aarão. 2010. Ditadura, anistia e reconciliação. *Revista Estudos Históricos* 23(45): 171–186.

Reis, Daniel Aarão. 2014. *Ditadura E Democracia No Brasil.* Rio de Janeiro: Zahar.

Ricoeur, Paul. 2006. *Memory, History, Forgetting.* Translated by Kathleen Blamey and David Pellauer. Chicago, IL: University of Chicago Press.

Rollemberg, Denise. 2006. Esquecimento das memórias. In Joao Roberto Martins Filho, ed., *O golpe de 1964 e o regime militar.* Sao Carlos: UFSCar, 81–91.

Schneider, Nina. 2011. Breaking the 'Silence' of the Military Regime: New Politics of Memory in Brazil. *Bulletin of Latin American Research* 30(2): 198–212.

Schneider, Nina, and de Almeida, GiseleIecker. 2018. The Brazilian National Truth Commission (2012–2014) as a State-Commissioned History Project. In Berber Bevernage and Nico Wouters (eds), *The Palgrave Handbook of State-Sponsored History after 1945.* London: Palgrave, 637–652.

Southgate, Beverley. 2007. Memories into Something New: Histories for the Future (Future Work of Historians). *Rethinking History* 11(2): 187–199.

Teitel, Ruti G. 1995. How are the New Democracies of the Southern Cone Dealing with the Legacy of Past Human Rights Abuses? In Neil J. Kritz (ed.), *Transitional Justice: How Emerging Democracies Reckon with Former Regimes,* Vol. I: *General Consideration.* Washington, DC: United States Institute of Peace, 146–154.

Teitel, Ruti G. 2000. *Transitional Justice.* New York: Oxford University Press.

Teitel, Ruti G. 2003. Transitional Justice Genealogy. *Harvard Human Rights Journal,* 16: 69.

Teitel, Ruti G. 2014. *Globalizing Transitional Justice: Contemporary Essays.* New York: Oxford University Press.

Torpey, John. 2003. *Politics and the Past: On Repairing Historical Injustices.* Lanham, MD: Rowman & Littlefield.

Torpey, John. 2006. *Making Whole What Has Been Smashed: On Reparations Politics.* Cambridge, MA: Harvard University Press.

Torpey, John. 2009. An Avalanche of History: The Collapse of the Future and the Rise of Reparations Politics. In Manfred Berg and Bernd Schaefer (eds), *Historical Justice in International Perspective: How Societies Are Trying to Right the Wrongs of the Past.* New York: Cambridge University Press, 21–38.

Verdeja, Ernesto. 2009. *Unchopping a Tree: Reconciliation in the Aftermath of Political Violence.* Philadelphia: Temple University Press.

Weiffen, Brigitte. 2012. From Domestic to International Instruments for Dealing with a Violent Past: Causes, Concomitants and Consequences for Democratic Transitions. In Aleida Assmann and Linda Shortt (eds), *Memory and Political Change.* New York: Palgrave Macmillan, 89–111.

Weyeneth, Robert R. 2001. The Power of Apology and the Process of Historical Reconciliation. *Public Historian* 23(3): 9.

White, Hayden. 1985. *Metahistory: The Historical Imagination in Nineteenth-Century Europe*. Baltimore, MD: The Johns Hopkins University Press.

Wilson, Richard. 2005. Judging History: The Historical Record of the International Criminal Tribunal for the Former Yugoslavia. *Human Rights Quarterly* 27(3): 908–942.

Yale University, Committee to Establish Principles on Renaming. 2016. Report of the Committee to Establish Principles on Renaming. Retrieved 4 June 2017 http://president.yale.edu/sites/default/files/files/CEPR_FINAL_12-2-16.pdf.

Zeitlyn, David. 2015. Looking Forward, Looking Back. *History and Anthropology* 26(4): 381–407.

10 Ruins, resistance and pluritemporality in Palestine-Israel

Luisa Gandolfo

Introduction

In the short story, 'Tea Outdoors', by Oz Shelach, a group of 14-year-old seniors are brewing tea in a forest park. As they 'learn to taste the land' by boiling the *za'atar* (thyme) sprigs that they picked on their hike, their journey gains a sensory layer as they learn the names of the herbs, as well as their flavours and aromas. The students' senses are restricted, however, and while they savour the taste and scent, Shelach broadens the vista to reveal that their surroundings are temporally divided. While the students are pouring and sipping in the now, it is a present blanketed with soil 'soaked with blood, where vines stretch out over ruins and persist like the claims – big sweet green grapes, rich in seeds, tall fig trees – of the farmers we drove away' (2003: 77).

At first, the landscape appears to the protagonists and the narrator to hold two levels: the bountiful forest of which the students are learning and the decaying past that nature is slowly reclaiming over time with her growing vines. However, on closer inspection a third layer emerges that unites the narratives, as the 'big sweet green grapes, rich in seeds, tall fig trees' are not only the claims of the exiled Palestinian farmers, but also the reason for the visit of the Young Guardians. As they move through the forest they engage with the surroundings slowly and carefully, learning, tasting, touching, and preserving. Their movements, we are told, denote a reverence that is lacking in the nameless 'indigenous peasants' who harvest the herbs indiscriminately. Yet despite their contemporary presence, the students fall between the layers of time as the narrator lifts the veil between the present and the past, and reverses the spectrality of the site. As the students become the 'ghosts that haunt the soil' (ibid.), their embodied presence is reduced to ephemera in the broader history and narrative of the land.

In this chapter I consider time, materiality, and embodiment in three nested contexts. Beginning with the outer layer, the landscape, I shall reflect on the significance of the landscape in the context of the depopulated villages of the West Bank. Over time, the land in and surrounding the sites has altered both naturally and forcibly, and the flora and lines that mark the area speak as much to memory and history, as they do to nature. In turn, through this chapter, I look at the ways that the past and the present is negotiated through memory and movement, with a particular focus on the land surrounding the depopulated villages of Iqrit and Kufr Bir'im. The second layer guides the chapter into the ruins of the villages and draws on Shelach's depiction of the temporal layers to consider the intersection of memory, matter, and the extent to which the villages stand as *noeuds de mémoire* (Rothberg 2010), as opposed to *lieux de mémoire* (Nora 1989). Finally, the chapter focuses on embodiment and restoration as a form of resistance in Kufr Bir'im and Iqrit, and through the work of the Palestinian heritage organisation, RIWAQ. Drawing on data gathered over the course of two years in Tel Aviv, Jerusalem, Jenin, Ramallah, Bethlehem, Haifa and Acre, the study is part of a broader, ongoing project that looks at the role of cultural activism and memory in peace-building in the region. The respondents were drawn from advocacy and heritage organisations and contacts were established with local community members through the organisations, who were asked to reflect on issues of displacement, resistance, and space.

Throughout the chapter, the overarching question concerns the ways that the animate and inanimate negotiate time, memory, and the occupation, as the sites transition from Palestinian villages to depopulated and contested sites. By challenging Halbwachs' observation that history is a 'dead memory', I argue that the 'organic' relation is sustained by engaging with the land through walking and restoration. Likewise, I shall consider the extent to which Huyssen's concept of 'the nostalgic lure' (2006) unfolds in the case of the restoration and the reutilisation of the ruins, and whether the 'authentic ruin' can be redefined in the context of ongoing territorial contestation.

Foundations: landscapes and the land

> Tree trunk against tree trunk,
> stone next to stone,
> thus the relationship takes form.
> We begin from here
> though doubts sometimes assail us.
> Sometimes we'll miss one another,
> but we won't be afraid to look behind.
> (Al-Qasim, 1987)

In Palestine-Israel, the 'land' stands as a cultural symbol; a focal point for the commemoration of loss; a strategic area that can be fought for, won, lost, and perhaps recaptured; she is the mother, the beloved, and the virgin (Sherwell 2003: 132); a tangible piece that can be gathered into bottles and pockets, and carried beyond borders as a reminder of home; it is a right; a nexus of mourning and resistance, and the bearer of two names, at times singular, at others divided by an oblique or united by a hyphen. Given the encompassing nature of the 'land' in this context of contestation, is it possible to disentangle the 'land' from the 'landscape'? Moreover, would attempts to distinguish between the two concepts diminish the political significance of the latter? Perhaps a way forward would be to consider the commonalities shared by the 'land' and the 'landscape' as memory practices and sites of commemorative activism. As a concept that arches over the region, the 'land' has presented a focal point for political activism, remembrance, and commemoration since Israel declared independence in 1948. As the decades passed and the land has been annexed, communities have been displaced, replaced, and exiled, as population flows prompt new towns to grow and old ones to be erased or renamed. The loss of the land is marked in a number of ways at a grassroots level, including through annual days of commemoration, such as Nakba Day, Naksa Day, and Land Day.

The first, Nakba Day, marks the 'catastrophe' (*an-Nakbah*), or, the loss of the land on 15 May, 24 hours after the Israeli Day of Independence (*Yom Ha'atzmaut*). The proximity of the dates renders *an-Nakbah* a counterpoint to *Yom Ha'atzmaut*, as for the latter celebratory events are held, while the former is marked by vigils, protests, lectures, and film screenings, occasions that engage multiple generations in a 'cottage industry of commemoration' (Allan 2007: 273) that draws on the cultural medium to reflect on loss, exile, and absence (Gandolfo 2015: 188). While the day's events vary according to the vicissitudes of conflict and proximity (Farsoun and Aruri 2006: 172; Jayyusi 2007: 131 n3; Masalha 2012: 8), the idea of the land merges with the body and the separation is experienced emotionally, as well as physically, as Saloul observes in his analysis of the song *Mawaal* ('Melody'), by Shafiq Kabha (1989), which alludes to a body 'that is torn into two halves / a living one and another that lived / and the living half is left for pain and suffering' (Kabha 1989 [Saloul 2008: 5]). While themes of physical and emotional pain are visible in Palestinian art, including the works of Sliman Mansour, Ismail Shammout, Emily Jacir, and Tamam al-Akhal; poetry, through writers such as Fadwa Tuqan, Dana Dajani, Mahmoud Darwish, and Samih Al-Qasim, and literature, encompassing Liana Badr, Ghassan Kanafani, Susan Abulhawa, and Sayed Kashua, the understanding of the land goes beyond artistic and literary practices, shaping the perception of space and place, and how belonging and absence is embodied on the sites (Abufarha 2008: 343).

The second, Naksa Day, marks the 'setback' incurred by the Six-Day War in 1967. Annually observed on 5 June, the war brought an unanticipated loss for not only the Palestinians, but also the participating forces of Egypt, Jordan, and Syria. By the end of the week, Israel had expanded its territory three times over (Monem et al. 2013: 118) as Israel occupied the West Bank, Golan Heights, the Gaza Strip, and the Sinai desert (Salibi 1993, 221). While the territorial loss was substantial, the displacement was more so, as 300,000 Palestinians (a quarter of the population of Gaza and the West Bank) were forcibly displaced (Masalha 2001: 67 n103) and sought refuge in Egypt, Jordan, and Syria. Of that number, 145,000 were UNRWA refugees who had been doubly displaced, having previously left their homes following the 1948 War (McDowall 1990, 84). The villages, now depopulated, joined the 400 villages and 11 cities that had lost their residents, names, and with time, their presence (Kadman 2015, 1). At this point the landscape gains a strategic significance as the perspective is transferred from the cartographical to the viewpoint of the individual gazing over the landscape. Looking towards the horizon, the dwellings – here erased, there renamed or rebuilt – establish new lines that are physically imposed, as in the case of ad hoc settlements surrounded by guards and barbed-wire (Gandolfo 2016a: 6), or unofficially so, through the shift in occupants and the varying degrees of access accorded to the traveller.

While the ruins could pique curiosity, the perception of their demise or absence – a term that I shall use to encompass renaming – is not uniform, as maps, history books, and *in situ* information boards determine the visitor's understanding of the sites. If, for example, the visitor is unaware of the sites, it is possible that the significance of the ruins will be less visible as the visitor focuses their attention on the beauty of the landscape or the conversation of their fellow travellers. As their conscious perception is diminished, the visitor could then,

> fail to notice what is right in front of [them] [...] Visibility, in other words, is thus a function of not just strictly physical factors such as the object's size or distance from the viewer, but also the extent to which he or she actually attends to it.
>
> (Zerubavel 2015: 1–2)

When the territorial narratives are contested, as in the case of the depopulated sites in Palestine-Israel, the question becomes whether a site can be seen the same way by everyone, or whether the gaze can be subjective according to what is known and unknown. Extending this further, we can consider whether Zerubavel's thoughts on collective ignoring can be applied, since the act of noticing and ignoring does not occur in an isolated context, but rather is a component of socialisation that is practised as a collective, whether in small groups – such as friend and family networks – or larger ones, such as communities or institutions (Zerubavel 2015: 9). When placed in the context of territorial and narrative contestation, the act of collective ignoring becomes part of the nation-building exercise as alternative interpretations of the history of the sites, and the sites themselves, become part of the narrative of the land, as well as the physical landscape itself.

However, the act of collective ignoring requires a framework in which to be enacted and time encompasses the key points that shape acts of remembrance, forgetting, and ignoring. As a reference point, time can be spliced and edited to form a collective narrative; it marks the points of heightened remembrance through designated dates of collective commemoration; it also signals the duration of the occupation and the need to negotiate a respite, whether at a state level through the peace process, or at a local level through grassroots peace initiatives. In certain cases, such as Rwanda (Brehm and Fox

2017), a pause is needed before addressing the trauma of the genocide (also see Benda, Chapter 8 in this volume). In Palestine-Israel, time follows an alternative path, aligning to Prager's understanding of a past lived 'as *if*' it was present, an approach that stops time, rendering it 'frozen', 'unyielding', and potentially threatening, as memory 'close[s] off all the gateways to the senses' (2006: 235). In turn, just as no two conflicts are the same, there is equally a lack of a definitive means by which to chart the negotiation of time after a conflict, as the nuances of the traumatic events, as well as how the actors address the events in the aftermath, determine whether time is stopped, paused, or continued. However, how a conflict is recorded, taught, or spoken about can reveal both the standpoint of the narrator, as well as the splicing in time that occurs through school textbooks and historical accounts.

In the context of the villages that were depopulated between 1948 and 1967, this is perceptible in a 1969 speech by the Israeli military leader and politician, Moshe Dayan. As he describes the transition of the land, he demonstrates the ways that the Palestinian landscape becomes subject to, as well as a site of, nation-building,

> Jewish villages were built in the place of Arab villages. You do not even know the names of these villages, and I don't blame you, since these geography books no longer exist. Not only the books do not exist – the Arab villages are not there either. Nahalal arose in the place of Mahlul, Gvat in the place of Jibta, Sarid in the place of Haneifa, and Kfar-Yehoshua in the place of Tel-Shaman. There is not one single place built in this country that did not have a former Arab population.
>
> (*Ha'Aretz*, 4 April 1969 [Massad 2006: 170])

It is noteworthy that Dayan does not 'blame' the public for their lack of awareness, indicating that their inability to recognise the sites is a by-product of the omission of the names of the sites and their inhabitants from the textbooks. Returning to the earlier question regarding Zerubavel's concept of collective ignoring, Dayan's speech indicates an effort to ensure that awareness is hindered, a result that is evocative both of Ram's three tiers of the 'regime of forgetting' (2009: 370) and Nora's reflection on the exclusion of an event, which makes it a *lieu de mémoire* (1989: 22). This could be taken a step further by locating both the ruins, as well as the days of commemoration, as *lieux de mémoire*. As Masalha argues, Nakba Day and Land Day are not only anniversaries, but junctures of 'strikes, demonstrations, defiance and resistance' (2012: 4) and 'the most important days on the Palestinian popular calendar' (ibid.). Occurring on specific dates, the days provide opportunities for commemorative activism that raise a paradox, as the dates are both timeless (Nora, 1989: 8), yet also guided by an awareness of time passing. As occasions that 'block the work of forgetting' (Nora 1989: 19), the commemorations held on the sites provide acts of resistance not only in terms of the embodied presence of the visitors, but also through the invocation of the former place names. These acts might not 'stop time' (ibid.), but they present a 'material, symbolic, and functional' (ibid.) means to remember the sites as they once were, through digital archives and the tours that take place through the year.

The final day, Land Day, has been commemorated annually on 30 March since 1976, though its roots lie 12 years before, when construction of the city of Karmiel in northern Israel, began. After several stages of land expropriation, the villages of Sakhnin, 'Arabyeh, and Deir Hanna were put forward in 1976 (Nakhleh 2011: 84). That March, the residents engaged in strikes and protests, but by the end of the month six Palestinian citizens were killed, 50 injured, and around 300 arrested (ibid.). The fusion of the body and the land, once more revealed, is set apart from the broader readings of experiencing the landscape (Merleau-Ponty 1965; Wylie 2009), as the loss of land and the redrawing of the landscape is marked by political events, such as the redrawing of boundaries and shifts in accessibility, rather than the natural passage of time. As Yiftachel notes, Land Day (this can be extended to Nakba Day and Naksa Day, too) represents an 'intimate intertwining between the geographical process – the discourses, development, and struggles over the land – and the construction of Palestinian symbols and identity' (2006: 69). To return to the question posed earlier, regarding whether the frame of landscape can be perceived equally by all visitors, the answer would be negative, since the way that we view an area is informed by memory and a prior awareness that shapes the way that we view ruins and the symbols that are attributed to the sites (Zerubavel 2015: 52; DeSilvey and Edensor 2012: 471). To this can be added Merleau-Ponty's argument regarding how the landscape is perceived, an action that is shaped first by the adage that 'to perceive is to remember' (1965: 19), and latterly, the reminder that before the memory can evoke the history of the site, the information must already be known in order to connect with the present vision (ibid.). For the commemorative activists, the ruins of the depopulated villages are perceived as sites of former homes, shops, and communal hubs. For visitors who are unaware of their relatively recent decline, they could be seen as sites that pre-date *an-Nakbah* and the events that followed. In turn, the unsuspecting visitor receives the site with a nostalgia that lacks awareness of the political and territorial contestation.

Constructs: walking through time

> The climbing body demands effort; it is under continuous tension. It is an aid to thought in the pursuit of examination: pushing on a little further, a little higher.
>
> (Gros 2015: 22)

Walking through the ruins, the visitors, perhaps consciously or subconsciously, negotiate 'the pluritemporality of the ruin, the convergence of material and personal memory, and the capacity of alternative, sensual engagements with the past' (DeSilvey and Edensor 2012, 471). While DeSilvey and Edensor locate the concept of 'pluritemporality' in the context of ruins, broader definitions of the term draw attention to not only the layering, but also the alternative interpretation of time. For Nowotny, pluritemporaility extends social time (1992: 422) beyond the 'time of physics or that of biology' (ibid.), allowing it to be subjectively and collectively interpreted according to the context. When located in conflicted societies, or communities that are in the post-conflict phase, social time can be affected by the duration of chronological time it takes for acknowledgement, justice, and/or reconciliation to be negotiated and achieved, as reflected elsewhere in this volume by van Roekel (Chapter 5) and Igreja (Chapter 7), among others.

Locating pluritemporality in Palestine-Israel, DeSilvey and Edensor's interpretation highlights the significance of the different stages of ruin in the negotiation of time, conflict, and narrating conflict. In Iqrit and Kufr Bir'im, the ruins are visible, but at locations where the rubble does not clearly denote an inhabited past, their decline has varied in terms of time, the process of depopulation, and the degree of deterioration that followed (Gandolfo 2016b). In Kufr Bir'im, the depopulation was prolonged by court appeals and displacement. For the inhabitants of other locales, such as Deir Yassin and Suba, the occupation was quick, and in the case of the former, resulted in many fatalities. Between April and October 1948, approximately 2,800 people were displaced from the four sites and 100 killed (Duffy 2001, 10). After the villagers were displaced from Deir Yassin, the houses were incorporated into the Kfar Shaul Mental Health Center in 1951 and during the intervening years, the remaining buildings have become warehouses or family homes, and the cemetery is overgrown (Khalidi 1992: 292; Wiles 2014b). Most notable, however, is the limited information on what passed, as memorials to the victims are located in New York, East Jerusalem, and Glasgow, but not Deir Yassin itself.

Walking through the complex, the contrast between what occurred in 1948 and the prosaic contemporary surroundings evokes Said's observation that the conflict is marked by the 'layering of the landscape' (2000: 180). As the new buildings cover the former, Deir Yassin is simultaneously a 'place' and a 'site', the hospital presenting a 'place' that is experienced tangibly today, while the history of the location reminds the visitor of the trauma that places it 'between other places' (Trigg 2009: 89). While the layers can be perceived as conceptual – if one knows the history of the site, one can sense the layers – its materiality speaks to the pluritemporality as the visitor passes the chain-linked basketball court and happens upon large rocks tumbled atop of one another, a pile that could be dismissed as rubble, but on closer inspection gains historical context and becomes 'the stuff of time itself' (Ingold 2012: 439), a fragment of an alternative reading of the site. The transition from rubble to a fragment of history marks the shift in perception that is equally present in the distinction between a 'site' and a 'place' (Trigg 2009: 89), and between 'rubble' and 'ruins'. Key to this transition is the narrative, which provides an aesthetic layer that brings a historical weight to the material object. As Huyssen observes in the context of the post-World War II European cities, the remains were rubble, the 'bombings, after all, are not about producing ruins' (Huyssen 2006: 8). Yet through the mediums of photography and film, the rubble gained a depth that invited the viewer/visitor to see beyond the rough piles and imagine what once stood on the site.

In Kufr Bir'im, the distinction between rubble and ruins is sharper as the structures have not been completely levelled, nor have they been built over. Rather, the archways, walls, and angular tops still rise, though nature creeps up and through the stones, the grass growing long and the branches climbing over tumbled rocks. The Church of Our Lady stands, topped with a bell added in 1975, after the original was lost in the early 1950s, and the cemetery remains, albeit subject to acts of vandalism. Depopulated in 1948, the 824 residents of Kufr Bir'im were displaced by Operation Hiram, which resulted in the occupation of the Upper Galilee region by the border with Lebanon (Zochrot 2010). Although the residents of Kufr Bir'im, and the surrounding towns and villages, were requested to leave for two weeks while the Israeli military conducted their operation, they were denied re-entry. Five years later, the residents were granted the right to return by the court, but the village was bombed by warplanes before they could return (ibid.). By 1972 the villagers were dispersed around Palestine-Israel, including the nearby town, al-Jesh, as well as Jordan and Lebanon, while the area was declared a closed military area that prohibited visitors entering the site. In 1973, after several years of legal negotiations, the villagers could visit the church and its cemetery (ibid.), yet despite this development, the ability to freely access and roam the site continues to be stymied.

After the displacement, the land around Kufr Bir'im was divided between the Jewish kibbutzim of Dovev and Baram and the remainder has become a national park and retreat (Gandolfo 2016b: 8). Despite the absence of residents, the ruins continue to provide a platform for embodied resistance that takes form through sit-ins, camp-ins, and attempts to construct dwelling spaces on the site. Since 1982, *Al-Awda Hariket Abna' Kufr Birim Al-Taqodumeyon* (The Return of the Progressive Sons of Kufr Birim) have organised *in situ* protests to mark Land Day and, since 1984, an annual youth camp, the Roots Camp, (Ghantous 2014) has been held in the nearby depopulated village of Iqrit. In 2012 the camps adopted a fresh approach and that year – Kufr Bir'im followed Iqrit in 2013 – they transformed from temporary sites of visitation and remembrance to 'return camps' that involved not only the youth, but the broader returnee community (Wiles 2014a). The decision to remain marked a transition from a politically static stance to an embodied one, as 'we saw that rather than a reactionary position 'appealing' for return and 'waiting' for it, we needed to adopt tactics of direct action to implement it' (ibid.). In turn, since 2012, the villagers and their descendants have marked their right to remain through commemorative occasions, including Nakba Day and Land Day, educational and cultural events, and the construction of restrooms and bathrooms for activists remaining on the site.

While the activists channel peaceful resistance, their presence has been challenged by the Israeli military and in 2009 regional council officials from Merom HaGalil demolished the bathroom and restrooms (Zochrot 2010). Within the depopulated space, heritage presents a ground on which the politics of recognition is contested (Weiss 2007: 414) and as the legality of the denial of access is disputed by subsequent generations, the location presents a symbolic site from which to assert the right of return. That the depopulated site was chosen over a central, urban location indicates the significance of debris as symbol of memory and resistance (Gandolfo 2016b: 7). As the activists move around the sites, the space is reclaimed through memory practices that establish their claim to the village, and their presence, which lends an added dimension to Cresswell's interpretation of the politics of material movement (Cresswell 2010: 21). As well as being defined by distance, speed, and frequency, the physicality of walking the sites draws attention to the location and the issue of access. In Kufr Bir'im, and other sites in Palestine-Israel, this is further defined by questions of 'who walks where?' and 'who has access to the locations?', as the opportunity to visit the site is conditionally granted, depending on who is visiting and when.

While the discourse on ruins is often tied to a consigned past – Roman ruins, eighteenth-century ruins, ruins from the early twentieth century – the depopulated villages are tied to ongoing issues of dispossession, rights, and exile. In certain cases, time is disrupted by conscious breaks that signal omissions from the nations' historical consciousness, for example, in the case of countries negotiating historical narratives following the fall of the Soviet Union. In Palestine-Israel, however, time flows in defiance of the natural (or political) junctures – death, displacement, exile – to form a seamless, long narrative that adds events as time passes. In turn, the past becomes accessible to all and in the context of villages, the memory of the site is carried by 'bearers' (Assmann 2011: 25), who impart the information through the camps, commemorations, digital archives, and walking tours. In doing so, the 'bearers' evoke Nowotny's point that time and space are subject to 'a multitude or plurality of temporal-spatial relations' (1992: 436), who set the pace of remembrance. The pace of recollection can be swift and is continuous, unlike in the case of Rwanda, in which Brehm and Fox note that an amount of time is needed to pass before the event can be added to collective memory (2017: 117). Rather than taking a moment before reflecting, the occupation becomes a catalogue to which traumatic events are continuously added.

In the case of the latter, RIWAQ, a Palestinian heritage movement that restores historical village centres, the restoration is but one element of remembrance. Through the walking tours, the team are introduced to the descendants and older residents, who share oral narratives that detail everyday life in the villages, prior to their decline. As they record and share the narratives, the names trigger deeper recollections among other residents, as Lana Judeh notes,

> We walk with old men and women from the village and they narrate to us stories from their past life in the historic centre, where they lived and what activities they did. We record this and we use this information and bring back the original names of the places. Some of the places are deserted now, but once we bring back the name of the place it comes back to the collective memory of the families who lived in the village.[1]

In turn, commemorative activism takes place in three ways. First, through embodiment as the walkers move through the ruins with an awareness of the sites' history. Second, where possible, by gathering the names and minutiae of everyday life prior to the decline of the sites, as demonstrated by the *iNakba* initiative and RIWAQ's oral narratives archive. Third, through engagement with the ruins, whether it is through restoration or by inhabiting the site as a location of protest, return, or reflection, as evidenced through the work of the Palestinian artist collective, *Insiyab* (Gandolfo 2015: 190). In cases where the ruins are regenerated, we can question whether Huyssen's argument regarding the repurposing of ruins (2006: 9) is applicable in the context of the renovation and repurposing of Palestinian buildings. Moreover, can restoration be considered an act of resistance against narrative erasure?

While there has been a strong theoretical analysis of ruins (Huyssen 2006; Weiss 2007; Trigg 2009; DeSilvey and Edensor 2012; Edensor 2005a, 2005b), there has been a limited consideration of ruins in the context of Palestine-Israel, which holds its own nuances as a post-colonial site marked by contestation, occupation, and dispossession. The differences brought by the destruction of the Palestinian villages do not render the observations of these authors redundant – there are important lessons to be drawn from each. Nevertheless, the nostalgia that is induced by the ruins gains singularity in the cases of Kufr Bir'im and Iqrit (to name but a few), as the residents are not consigning the sites to history, nor are they engaging in 'resistant' acts of nostalgia. Rather, nostalgia is present, and perceptible in the submissions to the mapping app, *iNakba*, by visitors to the sites, as they share information and anecdotes passed down by family members who once lived or worked in the villages.

Launched in April 2014 by the Israeli advocacy organisation Zochrot, *iNakba* raises awareness of *an-Nakbah* by inviting users to locate villages that have been 'completely ruined, destroyed, obliterated [...] partially demolished, or remained standing but were depopulated and their residents expelled' (Zochrot 2014) on the GPS map and add images, short films, anecdotes, and records. Once uploaded, visitors can access the app while walking through the sites and explore the history behind the ruins, as well as see the village as it was prior to its decline. Central to the app is the ability to roam the area – the map only functions on the mobile – and as the walker is guided by the map, they engage with the sites in a different manner, as walking presents 'a way of thinking and knowing [...] the walker is *thinking in movement*' (Ingold 2010: S135). At the same time, the data afforded by the app invites the walker to think about the ruins in the context of contestation and whether the walker is a descendant of members of the displaced community or not, they are drawn into the act of remembrance. While at this point the juxtaposition of 'sites' and 'memories' cues Nora, it is Assmann (2011: 44–45), and his reflection on Halbwachs' visit to Palestine in 1927 and 1939 (Iogna-Prat 2011: 823), who captures the nuances of the Palestinian villages effectively, as

> [E]ntire landscapes may serve as a medium for cultural memory. These are not so much accentuated by signs ('monuments') as raised to the status of signs, that is, they are *semioticized*. [...] Halbwachs set out to show through the commemorative landscape of Palestine that not only every epoch but also every group and every faith creates, in its own way, locations and monuments for its own particular memories.

Until this point, the layers have been projected onto the surface of the ruins and the landscape likewise. However, this could be taken further by considering whether the flow is one-way (from the walker to the landscape/ruins), or if, by virtue of being 'semioticized', the ruins and landscape are imbued with the capacity to trigger the walker. The possibility for the inanimate to engage with the walker in this manner would, Trigg writes, allow 'the ruin [to urge] us to approach testimony as an impossible demand, a break in the spatio-temporal presence' (2009: 96). However, while the ruin can 'urge' or prompt the walker to engage with the spatio-temporal dynamic, it can do so only if their attention is directed towards the ruin and, more importantly, informed of the history of the site. Without attention or information, the ruins risk returning to their status as 'rubble'.

The act of walking as a form of memorialisation has been replicated at locations where the buildings have been neither demolished nor absent, yet the buildings and walls of the city provide a performative backdrop that attempts to take the walkers back in time, adding an alternative take on the spatio-temporal dynamic. The two-day *Mapping Procession* was organised by the Palestinian artist, Khaled Hourani, for the 2014 Qalandiya International arts festival. The procession marked the first *intifada* (1987–1993) by guiding walkers through Ramallah with strategically placed works of embodied art by actors, circus performers, and artists, as well as installations, film, and music, including songs from the *intifada*. On the first day, 70 visitors joined the procession, which was accompanied by volunteers dressed in keffiyehs and clothes from the late-eighties, who distributed leaflets apprising the walkers of the *intifada* (Zwartsteen 2014). As the procession meandered through the city, the walkers 'intervene[d] and re-appropriate[d] public space and the citizen's relationship with it' (Qalandiya International 2014), bringing traffic to a standstill and subverting the flow of movement in the city.

The leaflets, sharing the date of the procession for the year 1988, called for resistance and for Hourani, their presence provides a commentary on the absence of sufficient political change as 'they can be distributed as if they were from today' (ibid.). On one level, the event provided an opportunity to inform and reflect on the *intifada*; on a deeper level, it questioned the status quo by reflecting on how far – and whether – the political situation has progressed with time. Ramallah, as a backdrop, a stage, and a location for the performance of memorialisation and resistance becomes reminiscent of the 'specious present' (Kern 2003, 82–83), a now that never quite evades the past, its fragments continuing to strike in the wake of the events, whether it is the Nakba, the Naksa, or the conflict and uprisings that have shaped the land since 1948. In turn, the perception of time shapes how memories are experienced and articulated, and space provides the backdrop that evokes both Nowotny (1992) and Prager (2006), as the present merges with the past and time is seen as malleable, as it brings elements of 1980s Ramallah into the present.

Movement: restoration and resistance

> The beauty of heritage is that you can create difference in the same, and sameness in the difference.[2]

Like memory, the sites discussed in this chapter are in a state of perceptive flux, chronologically layered on top of the other. Kufr Bir'im has three layers: as a Jewish village called Kfar Bar'am abandoned in the thirteenth century, as the Palestinian village, and currently divided between the kibbutzim Dovev and Baram. Iqrit was occupied by Crusaders, and Deir Yassin was part of a sixteenth century Ottoman settlement, Khirbet Ayn al-Tut (Gandolfo 2016b: 202). In turn, the villages are not ruins devoid of memory and connection, but sites sustained by collective memory that brings to mind Halbwachs' observation that 'the relationships established between stones and men are not so easily altered' (Jones 2011: 97–98). While Nora's *lieux de mémoire* draws on Halbwachs' work to address the multiple sites of memory, both tangible and intangible, Rothberg provides an alternative reading through the *noeuds de mémoire* (2010: 7). Responding to the limitations of Nora's concept, Rothberg proposes a less sweeping addition to the discourse on collective memory. While the former's *lieux* depend on collective memory and the community, Rothberg allows for an interpretation of memories and sites that promote 'rhizomatic networks of temporality' (ibid.) that allow memory practices to be revisited, redefined, and relocated. As the ruins are used in multiple ways by different parties to the conflict, the sites warm to Rothberg's *noeuds de mémoire* (2010: 7), as they are sustained by an ongoing process of memory work that is carried through a range of memory performances, including oral narratives, embodied actions, *in situ* story-telling, mapping, and restoration.

Founded in 1991 by the Palestinian author and architect Suad Amiry, RIWAQ's first project, *The Registry of Historic Buildings*, was compiled between 1994 and 2007 and comprehensively lists through three volumes 420 villages across the West Bank, Jerusalem, and Gaza. Comprising histories, maps, and images, the volumes provide both an archive through which to record the villages that had declined, as well as a foundation from which to restore buildings that would serve the communities that have remained in their vicinity. The objective, to save 50 villages that include 50,230 historic buildings, has resulted in a success rate of 87%, the measure of success depending on whether the buildings are not only restored (which 'is always 100%'[3]), but whether they are used, too. The work of RIWAQ is not restricted to rebuilding old sites, but extends to reviving the communal traditions by fusing the old with the new to create functional sites. The first step in the regeneration of the site is to engage the community and challenge their perceptions of the historical centres. As Khaldun Bshara observes, the sites have declined to the point that both the remaining inhabitants, and those who left the villages, view the sites dimly,

> According to *The Registry*, we have around 50,000 historic buildings, but half of them are in a very bad condition: they are deserted, crumbling, rubbish dumps. Usually they have a historic centre, but because it's fragmented from an ownership point of view, it becomes the place where people throw their garbage. If they don't need a chair or an old car, they send it there. Vegetation eventually grows, and the animals go there to die. It's a tragic place.[4]

While the communities of Suba, Kufr Bir'im, and Deir Yassin were displaced by the military, the reasons for the decline of the villages in *The Registry* vary according to economic and political factors. In instances where the communities declined due to the allure of the cities, the remaining inhabitants, often old or poor, entered into a cycle of decline: the village became more impoverished, in certain cases a drug culture emerged, and the villages and villagers, were stigmatised as 'unsafe',

> It becomes a vicious circle of desertion. If there isn't a sewer system, electricity, or water, people will leave. Who isn't leaving the historic centres are the poor, because they cannot afford to get out. This adds to the layers of deprivation as old people are stigmatised for living there. RIWAQ is trying to end this stigma. Living in a historic centre is a privilege and it's a dream for many people.[5]

Responding to this, the restored buildings work on two levels: first, they provide a communal centre that is facilitated through the implementation of the *al-Oneh* tradition of mutual assistance; second, once restored, they draw former and new residents back to the area. *Al-Oneh* draws on the traditional mechanisms of communal cooperation and fuses it with fresh approaches that have allowed community members to engage not out of necessity, but voluntarily. The result is a regeneration of the community as a whole as the initiative ripples out,

> Our Palestine was built slowly, but the people built it themselves, together. We intended to revive this concept to help each other. We started the housing project in the Qalqilyah area and we provided the materials, and the people provided the labour. When we did one house it meant that we did two, because we weren't paying for the labour.
>
> If I'm an electrician, then I can do the electricity for different houses; if you're a plasterer or a renderer, then you can render other houses, not just your own. In exchange for electricity you will do something for others, because you are not paying money – the money is avoided in this system.[6]

With time, the sites that had been in decline showed signs of promise. However, while for the residents and former residents the promise of a 'new' centre evoked clean lines and white buildings, the result differed, as RIWAQ aimed for historical accuracy over aesthetics. If Kern's interpretation of the 'specious present' is applied in the context of restoration, then it can be seen in RIWAQ's efforts to sustain the past in the present, a past that is not ossified, but travelling alongside the now. As Bshara notes, there is a need to observe the simultaneity in reconstruction,

> It's dangerous to think about the past as static and that we can go back to that moment. If we do that, we claim that the past is there and that we can go back and capture it in one shot. We try to respect the authentic material and upgrade it, from wherever it comes from: The Ottomans, the Mandate period, or even before.

To do this, the RIWAQ team takes the residents back to the records and structures the restoration according to the details held in The Registry, including the materials and cultural references. To not do so would be to 'falsify history', an act that is viewed as 'dangerous' since it omits historical narratives that have shaped the region over the centuries. Bshara's caution is one that resonates not only in the context of the historical villages, but the depopulated ones, too: 'It's dangerous to be in a kibbutz and not know what's new and what's old. And, it's dangerous to change the inscription on the wall and take something away because it's in Arabic or it's in Hebrew.' The designs and inscriptions, while unrecognisable to contemporary residents, lie at the heart of the need for recognition, including of pasts that might seem incongruous with the present.

The fusion of the past and the present brings back Huyssen's argument regarding the repurposing of ruins (2006: 9), and whether the question of authenticity is applicable in the context of the restored historical centres in Palestine. To an extent, there are parallels between the restoration and repurposing of old buildings in the villages and the ruins that are restored and used as venues for cultural events elsewhere in the world. When successful, that is, 'restored and used', the centres provide a communal site that initiates community engagement from the start of the restoration process to the end. Where they differ is through RIWAQ's faithfulness to the past, despite local ambivalence. In doing so, the sites are transformed from 'rubbish dumps' into active memory sites that speak simultaneously to the past, the present, and a future in which the villages are revived as residents return from the urban centres, whether intermittently, as vacation homes, or permanently. In doing so, the restoration 'jump-starts' time in the Pragerian sense, as the restoration goes beyond the repair of, for example, a long-collapsed roof and assists in the communal recovery from a prolonged period of 'entrenched psychic commitment to the stoppage of time' (2006, 238). That is, while the sites languished as markers of decline and loss, the restoration restores them to their past glory, as well as making them functional in the present, fusing past aesthetics with contemporary practicality.

The second point related to Huyssen's reflection on ruins considered whether restoration could be defined as an act of resistance against narrative erasure. During our conversation, Bshara was clear that restoration should not be viewed as a form of resistance. While he perceived a link between heritage and identity, he believed that 'our identity cannot be a resistance to their identity.' Rather, heritage would be defined not through direct political resistance ('we are sick of saying that we are "Political", with a large "P"'), but rather through 'the mode of life, caring, collecting and the collective. [Heritage] is the incubator of collective memory, rather than the memory itself'. In turn, renovation holds a symbolic value, as 'through renovation people try to recreate a past that is material',[7] and in doing so, their actions might be perceived as small acts of personal, or communal, resistance.

Conclusion

In *Picnic Grounds*, Shelach guides the reader through a myriad of landscapes, cityscapes, and memoryscapes. Each scape is layered as the narrator or a fellow protagonist lifts the veil between the present and the past. Through this chapter I have unpacked the landscape, the land, and the ruins from their nesting points and demonstrated the ways that individuals and groups engage with the temporal layers of the sites, depending on their perspective and level of historical and political awareness. The question remains, however, whether the land can be disentangled from the landscape, and if in the context of Palestine, such a severance would be detrimental to the significance of the latter. Assmann's observation that landscapes become 'semioticized' draws attention to the connected symbolism between the land and the landscape. While the predominance of the former is visible in the broader marking of the Nakba, Naksa, and Land Days, the symbolic significance of the landscape is stronger once the observers are *in situ*.

Similarly, while RIWAQ focuses on the rejuvenation of the villages, the locales are not divorced from the landscape that lies beyond the village boundaries. As Lana Judeh observes,

> It's not only about the built-up environment, but also the natural environment around it. [RIWAQ] maps the villages from an environmental perspective, from water drainage to fertile land, scenic views, and culture. Everything relates to the surrounding environment and the landscape of the village, which helps us to build a strategy for the historic centres.[8]

While the argument for extending the definition of heritage work and restoration as a form of resistance is valid, there remains a case for viewing regeneration and remembrance among the ruins as a form of activism. Reading the work of RIWAQ, and other cultural organisations who engage with the depopulated sites, through the lens of Ram's 'regime of forgetting', there are commonalities that hold the potential for resistance against the erasure of the history of the sites and their inhabitants. In particular, 'redemarcation' responds to the 'physical forgetting' that occurs when the Palestinian sites ('houses, neighbourhoods, villages and other landscape objects' (Ram 2011: 98)) are removed. The demolition of the sites, Ram argues, thwarted the residents' ability to reclaim them, while a new narrative created a 'memory hole' among the Israeli population (2011: 100). The work of Al-Awda, RIWAQ, and Zochrot functions within the extended definition of resistance, using embodiment, technology, art, and restoration to fill the 'memory hole' and provoke awareness of the depopulated sites. Within this definition, the archive becomes a site of resistance that fills the narrative spaces, contests alternative readings of the landscape, and sustains the 'specious present' at the sites.

When viewed as a relationship between the landscape and the self (Wylie 2009: 278), the land and the landscape become inextricably linked. As Wylie writes, our connection with the landscape can be understood in terms of 'ramifying bodily engagements, encounters and inhabitations' (ibid.) that occur amidst a 'living tapestry of practices, imaginations, emergences and erasures' (2009: 282). Once located in the context of Palestine-Israel, the dynamic between the self as the landscape, and the self on the landscape, becomes one, for as long as both are subject to contestation, the passing of time will not diminish the connection between (and to) the land, the landscape, and the depopulated sites.

Notes

1 Interview by the author conducted with Lana Judeh, architect and project coordinator at RIWAQ. Ramallah, Palestine. 14 December 2014.

2 Interview by the author conducted with Khaldun Bshara, director of RIWAQ. Ramallah, Palestine. 14 December 2014.

3 Ibid.
4 Ibid.
5 Ibid.
6 Ibid.
7 Ibid.
8 Interview by the author conducted with Lana Judeh. Ramallah, Palestine. 14 December 2014.

References

Abufarha, Nasser. 2008. 'Land of Symbols: Cactus, Poppies, Orange and Olive Trees in Palestine'. *Identities: Global Studies in Culture and Power* 15(3): 343–368. doi:10.1080/10702890802073274

Allan, Diana K. 2007. 'The Politics of Witness: Remembering and Forgetting 1948 in Shatila Camp'. In *Nakba: Palestine, 1948, and the Claims of Memory*, edited by Ahmad H. Sa'di and Lila Abu-Lughod. New York: Columbia University Press, 253–285.

Al-Qasim, Samih. 1987. 'From: After the Apocalypse'. In *Modern Arabic Poetry: An Anthology*, edited by Salma Khadra Jayyusi, 381–384. New York: Columbia University Press.

Aly, Abdel Monem Said, Shai Feldman and Khalil Shikaki. 2013. *Arabs and Israelis: Conflict and Peacemaking in the Middle East*. London and New York: Palgrave Macmillan.

Assmann, Jan. 2011. *Cultural Memory and Early Civilization: Writing, Remembrance, and Political Imagination*. Cambridge: Cambridge University Press.

Brehm, Hollie Nyseth and Nicole Fox. 2017. 'Narrating Genocide: Time, Memory, and Blame', *Sociological Forum* 32(*1*): 116–137. doi:10.1111/socf.12319

Cresswell, Tim. 2010. 'Towards a Politics of Mobility'. *Environment and Planning D: Society and Space* 28: 17–31. doi:10.1068/d11407

DeSilvey, Caitlin and Tim Edensor. 2012. 'Reckoning with Ruins'. *Progress in Human Geography* 37(4): 465–485. doi:10.1177/0309132512462271

Duffy, Terence M. 2001. 'Museums of "Human Suffering" and the Struggle for Human Rights'. *Museum International* 53(1): 10–16. doi:10.1111/1468-0033.00292

Edensor, Tim. 2005a. 'The Ghosts of Industrial Ruins: Ordering and Disordering Memory in Excessive Space'. *Environment and Planning D: Society and Space* 23(6): 829–849. doi:10.1068/d58j

Edensor, Tim. 2005b. *Industrial Ruins: Spaces, Aesthetics and Materiality*. Oxford: Berg.

Farsoun, Samih K. and Naseer Aruri. 2006. *Palestine and the Palestinians: A Social and Political History*. Boulder, CO: Westview Press.

Gandolfo, Luisa. 2015. '(Re)constructing Utopia: Remembering and Forgetting in Palestinian and Israeli Art'. *Third Text* 29(3): 184–194. doi:10.1080/09528822.2015.1102400

Gandolfo, Luisa. 2016a. 'Transactions, space and Otherness: Borders and Boundaries in Palestine–Israel'. *Journal of Cultural Geography* 33(3): 253–274. doi:10.1080/08873631.2016.1184433

Gandolfo, Luisa. 2016b. 'Material Mnemonics and Competing Narratives in Palestine-Israel'. *Social Semiotics* 26(6): 1–16. doi:10.1080/10350330.2016.1187877

Gros, Frédéric. 2015. *A Philosophy of Walking*. London: Verso.

Huyssen, Andreas. Spring 2006. 'Nostalgia for Ruins'. *Grey Room*, 23–21.
Ingold, Tim. 2010. 'Footprints through the Weather-World: Walking, Breathing, Knowing'. *The Journal of the Royal Anthropological Institute* 16: S121–S139. doi:10.1111/j.1467–9655.2010.01613.x
Ingold, Tim. 2012. 'Towards an Ecology of Materials'. *Annu. Rev. Anthropol.* 41:427–442.
Iogna-Prat, Dominique. March 2011. 'Maurice Halbwachs ou la mnémotopie: "Textes topographiques" et inscription spatiale de la mémoire'. *Annales. Histoire, Sciences Sociales* 66: 821–837.
Jayyusi, Lena. 2007. 'Iterability, Cumulativity, and Presence: The Relational Figures of Palestinian Memory'. In *Nakba: Palestine, 1948, and the Claims of Memory*, edited by Ahmad H. Sa'di and Lila Abu-Lughod. New York: Columbia University Press, 107–135.
Jones, Paul. 2011. *The Sociology of Architecture: Constructing Identities*. Liverpool: Liverpool University Press.
Kadman, Noga. 2015. *Erased from Space and Consciousness: Israel and the Depopulated Palestinian Villages of 1948*. Bloomington, IN: Indiana University Press.
Kern, Stephen. 2003. *The Culture of Time and Space, 1880–1918*. Cambridge, MA: Harvard University Press.
Khalidi, Walid. 1992. *All That Remains: The Palestinian Villages Occupied and Depopulated by Israel in 1948*. Washington, DC: Institute for Palestine Studies.
Masalha, Nur. 2001. 'The Historical Roots of the Palestinian Refugee Question'. In *Palestinian Refugees: The Right of Return*, edited by Naseer Aruri, 36–71. London: Pluto Press.
Masalha, Nur. 2012. *Palestine Nakba: Decolonising History, Narrating the Subaltern, Reclaiming Memory*. London: Zed Books.
Massad, Joseph. 2006. *The Persistence of the Palestinian Question: Essays on Zionism and the Palestinians*. Abingdon: Routledge.
McDowall, David. 1990. *Palestine and Israel: The Uprising and Beyond*. London: I.B. Tauris.
Merleau-Ponty, Maurice. 1965. *Phenomenology of Perception*. London: Routledge & Kegan Paul.
Nakhleh, Khalil. 2011. '*Yawm al-Ard* (Land Day)'. In *The Palestinians in Israel: Readings in History, Politics and Society*, edited by Nadim N. Rouhana and Areej Sabbagh-Khoury. Haifa: Mada al-Carmel.
Nora, Pierre. Spring 1989. 'Between Memory and History: Les Lieux de Mémoire'. *Representations* 26: 7–24. doi:10.2307/2928520
Nowotny, Helga. 1992. 'Time and Social theory: Towards a Social Theory of Time', *Time & Society* 1(3): 421–454. doi:10.1177/0961463X92001003006
Prager, Jeffrey. 2006. 'Jump-Starting Timeliness: Trauma, Temporality and the Redressive Community'. In *Time and Memory*, edited by Jo Alyson Parker, Michael Crawford and Paul Harris, 229–245. Leiden: Koninklijke Brill.
Qalandiya International. 2014. 'Mapping Procession: Day 1'. N.d. *Qalandiya International*. Retrieved 30 January 2015. www.qalandiyainternational.org/2014/events/mapping-procession-day-1.
Ram, Uri. 2011. 'Ways of Forgetting: Israel and the Obliterated Memory of the Palestinian Nakba'. *Journal of Historical Sociology* 22(3): 366–395. doi:10.1111/j.1467–6443.2009.01354.x
Rothberg, Michael. 2010. 'Between Memory and Memory: From Lieux de mémoire to Noeuds de mémoire'. *Yale French Studies* 118/119: 3–12.

Said, Edward. 2000. 'Invention, Memory and Place'. *Critical Inquiry* 26(2): 175–192. doi:10.2307/1344120

Salibi, Kamal. 1993. *The Modern History of Jordan*. London: I.B. Tauris.

Saloul, Ihab. 2008. '"Performative Narrativity": Palestinian Identity and the Performance of Catastrophe'. *Cultural Analysis* 7: 5–39.

Shelach, Oz. 2003. *Picnic Grounds: A Novel in Fragments.* San Francisco, CA: City Light Books.

Sherwell, Tina. 2003. 'Imaging the Homeland: Gender and Palestinian National Discourses'. In *After Orientalism: Critical Entanglements, Productive Looks*, edited by Inge E. Boer, 123–145. Amsterdam: Rodopi.

Trigg, Dylan. 2009. 'The Place of Trauma: Memory, Hauntings, and the Temporality of Ruins'. *Memory Studies* 2(1): 87–101. doi:10.1177/1750698008097397

Weiss, Lindsay. 2007. 'Heritage-Making and Political Identity'. *Journal of Social Archaeology* 7(3): 413–431. doi:10.1177/1469605307081400.

Wiles, Rich. 2014a. 'The Evolution of Direct Action in the Struggle for Palestinian Return'. *Middle East Eye*, 17 May. Retrieved 13 January 2017. www.middleeasteye.net/essays/evolution-direct-action-struggle-palestinian-return-1640234501.

Wiles, Rich. 2014b. 'Remembering Deir Yassin'. *Al Jazeera*, 16 April. Retrieved 11 January 2017. www.aljazeera.com/indepth/inpictures/2014/04/pictures-remembering-deir-yassi-201441485916733186.html.

Wylie, John. July 2009. 'Landscape, Absence and the Geographies of Love'. *Transactions of the Institute of British Geographers* 34(3): 275–289. doi:10.1111/j.1475-5661.2009.00351.x

Yerushalmi, Yosef Hayim. 1996. *Zakhor: Jewish History and Jewish Memory.* Seattle: University of Washington Press.

Yiftachel, Oren. 2006. *Ethnocracy: Land and Identity Politics in Israel/Palestine.* Philadelphia: University of Pennsylvania Press.

Zerubavel, Eviatar. 2015. *Hidden in Plain Sight: The Social Structure of Irrelevance.* New York: Oxford University Press.

Zochrot. 2010. *Remembering Kufr Bir'im*. Tel Aviv, Jaffa: Zochrot.

Zochrot. 2014. 'iNakba: The Invisible Land'. *Zochrot*. Retrieved 13 January 2017. http://zochrot.org/en/keyword/45323.

Zwartsteen, Emily. 2014. 'Artistic Procession Reflects Palestinian Past'. *AURDIP*, 5 November 2014. Retrieved 14 January 2017. www.aurdip.fr/artistic-procession-reflects.html?lang=en.

11 Conclusion: Defusing time bombs

Towards an understanding of time and temporality in peacebuilding

Natascha Mueller-Hirth and Sandra Rios Oyola

Introduction

When we talk about societies that have faced mass violence, we wonder how people can ever 'move on' from this, we question how the past can be 'left behind', and we hope that future generations will heed the lessons from the past. In the narrower context of transitional justice theory and practice, the timing and duration of particular mechanisms are debated, while peacebuilding efforts need to achieve transformations over the short-, medium- and long-term, with the knowledge that unresolved issues can act as time bombs in new waves of conflict. Time permeates so many aspects of transitional and post-conflict societies, but we often take it for granted and rarely reflect on how it is understood and experienced and, consequently, how it shapes these societies. To do so can reveal neglected issues that are important for building sustainable peace; in other words, they can go some way towards understanding how to defuse time bombs. This has been the rationale for this volume.

The irony has not escaped us, the editors, that we have continued to use the terms 'transitional' and 'post-conflict' societies (including in the very title of this book), while trying to problematise the assumption of a clear break between violent past and peaceful present that is implicit in so much of the transitional justice and peacebuilding fields. Nancy Munn springs to mind here, when she writes that 'like all other discourses, those about time themselves take temporal form' (1992: 94). This brief conclusion will first highlight the main findings of the collection as a whole and will then discuss the scholarly contribution and policy implications of its focus on time and temporality.

Key findings

The chapters have explored key problematic dimensions of transitional and post-conflict societies through a time-sensitive lens. These dimensions are central to sustainable peace after mass violence and include reconciliation (Rosoux, Chapter 2), transitional justice institutions (Hinton, Chapter 3), peace processes (Rios Oyola, Chapter 4), victims and perpetrators (van

Roekel, Chapter 5), accountability (Igreja, Chapter 6), reparations (Mueller-Hirth, Chapter 7), intergenerational transmission (Benda, Chapter 8), historical narrative (Iecker de Almeida, Chapter 9), and the intersection between time, memory and place (Gandolfo, Chapter 10).

In each of these contributions, the use of a temporal approach has brought to the fore neglected issues that pertain to the case studies themselves. There are however some overarching findings that have emerged from the book as a whole. In what follows, we explore three key dimensions of the problem of time in societies emerging from conflict: the conflicts and negotiations of local and institutional temporalities; the relationships of power that underlie constructions and experiences of different temporalities; and the relationships between time and memory.

Local and institutional temporalities

There are differences, and frequently temporal conflicts, between local, or lived, temporalities and institutional, or official, temporalities. Despite the contextual differences between the countries and cases, this theme runs through all the chapters and helps to illuminate temporal clashes or conflicts in relation to reconciliation, inter- and transgenerational dynamics, accountability, reparations, collective memory, forgiveness, and so on. Before exploring this finding in greater detail, it is important to acknowledge that scalar frames such as 'local' are socially constructed; by using this terminology, we are referring to the lived experiences of time, peacebuilding and transitional justice by people who are directly affected by them.

Rosoux has argued this point in relation to the ability or willingness to reconcile, citing the experiences of a Rwandan survivor, Révérien Rurangwa, whose healing – like that all of survivors – does not adhere to the time frames envisaged by the national reconciliation process. As he puts it, 'How can we fix in advance a standard pace of healing?' This theme is picked up again and again throughout the volume. Healing, or 'moving on', is personal and happens at various paces and in non-linear ways, but it is also socially constructed and affected by wider structural issues and the specific experiences of past and, sometimes, present violence. Also in Rwanda, Benda's research on the Youth Connekt Dialogues (YCD) between children of perpetrators and children of victims describes the dialogues as 'atypical' reconciliation encounter – both groups can be seen to be victims of the *genocidaires*' crimes, although they have been affected differently. The YCD are an instance of a subaltern temporality that Benda contrasts with a rear-mirror, past-focused and exclusively survivors-centred temporality of the older generation.

Van Roekel's research in Argentina allows us to see that there are a range of different experiences of time that produce different understandings of local transitional justice trajectories. While the perpetrators in her study had more stable understandings of the past, violence, for them too, was a perpetual rather than a transient social condition, contrary to what a linear transitional justice temporality might assume.

We see this vast difference between local and institutional temporalities very clearly in Igreja's work too. Embodied, or local, accountability occurs at family and community level to resolve violations that have been forgotten, distorted or hidden by diverse social actors over time. When spirits take possession of a body, accountability for past serious violations occurs irrespective of the passage of chronological time and is not restricted to individual perpetrators, as official processes of accountability by transitional justice institutions do. As such, embodied accountability expands the temporal scope and temporal rhythms of liability and highlights 'the tense coexistence of past and present and the need of accountability to attain resolution'. It is noteworthy that the Mozambican case has been characterised by an absence of global or national accountability efforts for the civil war violations – Igreja importantly demonstrates that embodied accountability can contribute to justice in local terms and combat impunity.

When Rios Oyola analyses the peace plebiscite and public apologies in Colombia as a mechanism for accelerating forgiveness and the social peace process in order to catch up with the political peace process, she similarly draws attention to the different temporalities that are inherent to each. The social peace process is framed by the lived experiences of conflict, whereas the political peace process is framed by a teleological narrative and responds to international and national pressures. Rios Oyola's approach illustrates the negative effects of such 'accelerators' for the carefully achieved balance between individual and collective forgiveness in local communities affected by conflict.

While the Colombian peace plebiscite took place recently, in 2016, Mueller-Hirth's study on reparations in South Africa discusses a conflict that ended over 20 years ago. It is partly this long passage of time that explains why there is a dominant understanding in the political and public spheres that victims/survivors should long have moved on. By contrast, many of the survivors Mueller-Hirth interviewed were 'still waiting': for payment of reparations and the so-called struggle pension, but also for the promised transformative change. Uncertainties about waiting times and the bureaucratic process increased feelings of insecurity and powerlessness among people who are already marginalised due to historical, and present-day structural, violence. Again, in this case, victims' lived experiences of time clash with mainstream understandings and expectations about how long people 'should' take to move on.

Time and power

Temporal relations mirror power relations and power relations are made through the medium of time. The contributors to the book evidence the complex links of time with power and inequality empirically and in relation to specific issues faced by societies emerging from conflict.

For example, relationships of power can be found in the ability to control timings and set deadlines, as many of our contributors argue. Rosoux's work indicates that, in most cases, international actors dictate the timing of reconciliation efforts. Rios Oyola, in her analysis of the Colombian peace plebiscite, likewise demonstrates the ability of certain actors in transitional and post-conflict societies to influence, and accelerate, the temporality of peace processes. Relationships of power also characterise processes of forgiveness, as she shows in relation to the victims of the Bojayá massacre: different actors used the language of collective forgiveness in order to influence the vote. It is also possible to think about the ability to control timing as a strategy of power in relation to accountability: Igreja's research in Mozambique underscores how official mechanisms of accountability with their deadlines and statute of limitations impose a hierarchy of importance about the events that took place more recently. Conversely, embodied accountabilities do not present such hierarchies. When considering the institutions and mechanisms of transitional justice and peacebuilding from a temporal perspective, we can become alert not only to the power to impose deadlines but also the power to assert a particular temporality and transitional justice imaginary. Hinton, in his research on the Khmer Rouge Tribunal, suggests that international actors and local elites constitute themselves through the assertion of a transitional justice imaginary, which is normative, performative and productive, and premised on teleological 'transitional justice time'.

Control over timing and pace can render reconciliation processes 'too soon' for those affected by violence. As we have seen, healing practices themselves clearly are not neutral; they involve particular power relations (for example, when it is time to move on) and establish power hierarchies (who sets the pace and who is found lagging behind). While sometimes the imposition of peace processes can seem 'too soon' or 'too much', this book has also highlighted the disempowering effects of 'too slow' – or of 'not at all' – as has been the case in relation to reparations in South Africa. Mueller-Hirth's research shows the impact of the continued wait for reparations and transformation on victims/survivors of Apartheid-era human rights violations. Her focus on waiting as a less noticeable form of power clearly demonstrates the links between time and inequality: most of the victims/survivors were poor and marginalised, which meant that they were made to wait in the first place, could not afford to stop waiting, and the uncertainties of the waiting experience gave them less control over their lives and the future. The fact that the government continues to delay without entirely destroying hope further consolidates their fixed liminality.

Time and memory

It is often said that time heals all wounds. However, Rosoux (Chapter 2) explains that time alone is not enough to transform the meaning of the past. Instead, the lack of psychological closure can be found in the nature of the violence that was committed in the past. Her findings in the Franco-Algerian case imply that an asymmetrical war impedes the process of reconciliation and the transformation of memories through the mere passing of time. As Gandolfo (Chapter 10) puts it succinctly, 'the perception of time shapes how memories are experienced and articulated'. Gandolfo's research considers how ruins in Palestine-Israel are related to ongoing issues of occupation, dispossession, exile and camps; commemorative activism through walking tours acts as resistance to the natural passing of time, in which 'time is disrupted by conscious breaks that signal omissions from [a nation's] historical consciousness'.

Iecker de Almeida explores the 'un-doing' of Brazil's past as the creation and legitimation of preferred historical narratives that is consistent with a desired future. The creation of new historical narratives corresponds to an idea of the expected future for Brazil; in this narrative the dictatorial regime is overcome and Brazil appears as 'a country that promotes and upholds international human rights norms and discourse'. The notion of an envisioned future for a country contributes not only to transforming the memories of the unwanted past but it can deconstruct 'a legacy and memory thought to be fundamentally at odds with that vision'.

Van Roekel's chapter argues that Argentine victims of state repression build their memories under 'messy temporal orders', understanding in a different manner how past, present and future are connected. The victims/survivors she interviewed did not conceive of closure, or of trauma eventually being overcome; remembering was based on a spiral understanding of time and their traumatic pasts continuously intersected in the present. Her findings demonstrate that multiple co-existing lived experiences of time also produce multiple forms of memorialisation and multiple meanings of transitional justice. In a similar vein, Igreja conceives of local practices of conflict resolution as a way of bearing witness, which is enabled through forms of collective remembering embedded in diverse temporalities that are not bound by strict individual responsibility.

Scholarly relevance and policy implications

The aim of this book has been to think more explicitly about time when we study transitions and post-conflict societies, rather than to take it for granted or to ignore it altogether. We believe that this is theoretically important because time underpins all aspects of everyday life, it is both common-sensical and highly abstract (West-Pavlov 2012), and it has occupied sociologists ever since it became a discipline (Adam 1990). The wide range of issues in transitional societies within which time is imbued strongly justifies, we think, our theoretical approach.

But, given what is at stake in transitional and post-conflict societies, this has not been our primary aim. We began thinking about bringing together temporal analyses because we believe that they can help us to better understand the lived experiences of people in these societies, the gaps between transitional justice policies and their implementation, how to tackle the long-term challenges faced by societies after conflict and, ultimately, what contributes to sustainable peace.

As discussed above, the gap between local and institutional temporalities constitutes a shared finding across all the chapters in this volume. To be reliant on a particular conception of time, one that imagines societies to be moving forward in fairly linear ways without recognising that people's senses of time do not necessarily conform to this, can be detrimental to rebuilding relationships after conflict. To pay inadequate attention to people's experiences of time and their changing needs and demands can lead to a further marginalisation of people who are already marginalised during and after conflict. If the aim of transitional justice is to redress past wrongs and build peaceful futures, then it needs to be open to local and non-linear temporalities, and to victims' and perpetrators' experiences of lived time.

Indeed, as some of the chapters in this collection show, people affected by violence use alternative conceptions of time in order to cope with the long-terms effects of human rights violations – sometimes in addition to existing mechanisms, sometimes in lieu of them. This can be crucial for restoring relationships and building peace, as well as respecting people's own ways of coping with violence. Moreover, the acknowledgment of multiple temporalities has implications for justice. Local efforts of accountability, and their distinct ways of being in time, can provide justice and contribute to processes of bearing witness and remembering, particularly where global transitional justice processes have failed or have been absent.

This would also imply recognising, at institutional levels, that emotions, such as anger and guilt, or needs, for example for justice, might not develop in linear ways and can be shaped by socio-economic and political developments in post-conflict societies. Similarly, forgiveness or reconciliation do not necessarily develop sequentially until some complete and final state is reached. Research by the contributors to this volume argues, by contrast, that people's understandings of these terms not only differ from official or institutional definitions, but also dynamically change over time, often informed by their circumstances and contemporary experiences of violence and (in)security.

Consequently, transitional justice mechanisms and peacebuilding strategies should echo the different temporalities of people affected by violence. While it is true that official institutions are bounded by national and international deadlines as well as by political pressure and statutes of limitations, mechanisms of accountability and reparation should nonetheless seek to incorporate the diverse temporalities experienced by victims, particularly those alternative to the linear perspective of time. Paying attention to these different notions of temporality could, for example, complement or enhance current amnesty laws or reparation policies. In the words of Igreja in this volume, a conscious attention to time in transitional justice policies can help to 'attain accountability in ways that are also meaningful for the diversity of people victimised by an array of serious violations over time'.

This book has also highlighted that the timings and implementation of reparations and other transitional justice policies can lead to temporal conflicts and marginalisation. The logic of speed, the vulnerability of transitional societies, and international pressures all imply that processes and mechanisms need to be put in place very quickly; then, however, the implementation of reparations and victim support is frequently delayed or not delivered at all. The politics of waiting, as they play out in relation to these issues, can negatively shape people's engagement with the state and former enemies in the post-conflict settlement. No unrealistic promises should be made to victims about reparations or, at the very least, uncertainty about reparations and support should be reduced, since uncertainty can prevent people from exerting control over the future and from moving on from the past.

Moreover, to accelerate peace processes beyond the temporality of people's lived experiences can also be detrimental for local processes of reconciliation and can ultimately threaten the development of sustainable peace. Political peace processes usually favour the introduction of fast-paced liberal economic systems which might be at odds with the temporal logic of restitution and redistribution. However, for transitional justice to become transformative, it needs to tackle structural violence and longer-term social change and involve participation from and empowerment of victims. In other words, a transformative perspective will not only include the incorporation of development issues in transitional justice scenarios (see for example Duthie 2008; Waldorf 2012), but will also involve the recognition of victims' definition of conflict and projects of peacebuilding and reconciliation.

It is far better recognised that social and collective memories have the potential both to prevent a renewal of violence, and to derail pace processes and become fuel for time bombs in new waves of conflict. However, it is less well understood how past and future horizons of reconciliation projects can be balanced when countries 'go back in time and create different timelines' (Iecker de Almeida, Chapter 9) that can service its desired future. When societies 'undo' the past by making it seem as though it never happened, past, present and future are linked in complex non-linear ways. Rather than correcting past injustice, this kind of undoing might serve to legitimise particular political agendas or forms of injustice.

The research in this book shows that temporal analyses can be particularly useful for examining the challenges that mature post-conflict societies face a relatively long time after transition and for recognising the long-term requirements of victims, perpetrators and their children and children's children. Again, this is important because of how these unresolved issues can impact countries' longer-term prospects for peace and stability. The future-oriented vs. past-oriented policies of the transitional justice paradigm are no longer sufficient to explain the social and political transformations of transitional societies. Transitional justice should not work as a Janus-face prism where accountability is juxtaposed with prospective measures that seek to build democratic futures. Instead, the findings in this volume indicate that the messiness of the subjective and social experience of time, the multi-temporalities and pluri-temporalities, are part of the everyday life of people who have to live with the legacies of mass violence. The fluidity of their negotiations between past and present need to be recognised beyond providing an interesting anthropological vignette. As such, they should be operationalised in plans and policies of reconstruction, development, reparation and accountability. One way of achieving this operationalisation is through a broader and more flexible approach in the recovery of truth and historical memory that recognises victims' definitions of conflict instead of imposing particular time constructions in the definition of conflict. Similarly, programmes of reparations and retributive justice should avoid creating new forms of re-victimisation and re-traumatisation through politics of waiting.

We hope that this book serves to make a case for the need to engage with the category of time and that it manages to shed light on dimensions that are crucial for building sustainable peace. There are, of course, a whole range of issues that are central to transitional and post-conflict societies and that we have not addressed in this volume. For example, both editors have worked on gender issues in conflict and post-conflict contexts and are aware that the book does not explore, in any depth, the ways in which temporal experiences and temporal power relations are gendered in transitional societies and beyond. The reintegration of perpetrators and ex-combatants, and a deeper analysis of generational issues, such as the impact of transitional justice practices across different generations, are two other key areas we would have liked to have considered in relation to the objectives of this book. We hope that such concerns, and others, can be explored in future work that critically engages with time, temporality and power.

References

Adam, Barbara. 1990. *Time and Social Theory*. Cambridge: Polity.

Duthie, Roger. 2008. Toward a Development-Sensitive Approach to Transitional Justice. *The International Journal of Transitional Justice*, 2(3), 292–309.

Munn, Nancy D. 1992. The Cultural Anthropology of Time: A Critical Essay. *Annual Review of Anthropology*, 21(1), 93–123.

Waldorf, Lars. 2012. Anticipating the Past: Transitional Justice and Socio-Economic Wrongs. *Social & Legal Studies*, 21(2), 171–186.

West-Pavlov, Russell. 2012. *Temporalities*. Abingdon: Routledge.

Index

FSC
www.fsc.org
MIX
Papier | Fördert gute Waldnutzung
FSC® C083411

Kolibri 360 GmbH
Ferdinand-Jühlke-Straße 7
99095 Erfurt, Deutschland
produktsicherheit@kolibri360.de